A Story of Dreams, Doubts, and Triumph

Under The Midnight Sky

The Memoirs of

Geoffrey L Yoder

From Farm Boy to NASA Executive

Editor: Rachel Arterberry at Making A Way Writing Services
www.makingawaywriting.com

eBook ISBN: 979-8-9987404-0-4

Paperback: 979-8-9987404-1-1

Cover Photo Credit: Hubble Image of the Eagle Nebula Pillars of Creation

Photo Credit: NASA and ESA

Praise for Under the Midnight Sky

"With more than 16 years in the industry and 16 years at NASA, Geoff's story is not only of individual success and hard work, but also of NASA's transition to a new era of space exploration, in which he played many key leadership roles. He has accomplished what most of us come here hoping to do—move our mission—and America's space program—forward." *Former NASA Administrator, Retired Major General Charles F. Bolden Jr.*

"I had the honor of working with Geoff while I was at NASA. He was a constant supporter and advisor on how to navigate NASA and bring greater awareness on the importance of cybersecurity as part of mission delivery. I am forever grateful for his warm welcome!

What awaits you in these pages is not merely a collection of words but a pathway to your dreams, even dreams you didn't even know you had. As you stand at this threshold, know that you hold in your hands more than just a book—you hold possibility.

In life's journey, remember that your worth is determined not by others' judgments but by your own belief in yourself. The voice inside you that whispers "possible" when everyone else says "impossible" is the one worth heeding.

Yet self-belief doesn't mean isolation. Reaching out takes courage and wisdom—there is strength in seeking help when needed.

Stay curious. The moment we think we know everything is when we stop growing. Question, explore, wonder.

Finally, share what you learn. In teaching, we deepen our own understanding. In mentoring, we create ripples that extend far beyond ourselves.

This book is your companion on the path ahead. May it inspire you to believe, connect, question, and uplift others along the way.

Turn the page. Your story awaits." *Retired NASA Chief Information Officer, Renee Wynn*

I dedicate this book to my wife, son, daughter-in-law, and grandson, whose unwavering support has been invaluable.

I also extend my gratitude to the NASA leadership for their trust in my ability to lead various organizations.

A special appreciation to all the teachers and educators who tirelessly educate our youth, and to all readers seeking to reach their dreams, even when it seems impossible.

With best regards,

Geoff Yoder

CONTENTS

REFLECTIONS

As I sat in the hotel restaurant in Germany, enjoying my tea, I couldn't help but chuckle at the idea of a farm boy like me holding a senior executive role at NASA, leading discussions with 13 Space Agencies about exploration.

Unfortunately, I was never a great student. By the fourth grade, I did not understand why math was so important, so I didn't try, leading to a nearly failing grade. In seventh grade, a teacher insisted on saying my name incorrectly (in fact that I was pronouncing it wrong), which was demoralizing, causing me to have a poor attitude toward education. My high school algebra and Physics teachers regularly told me I wasn't trying. Finally, in my junior year, after seeing that I was not putting in any effort, a teacher stated that since I grew up on a farm, I would not amount to anything other than a farmer.

Those words stung me to my core and made me doubt myself. However, with the support of friends and family and, as you will learn, one coworker in particular who nearly took me by the hand, I learned that other people's opinions do not define me or determine my success. It was not easy, but through hard work and perseverance, I surpassed even my own expectations. It is like a crazy plot twist that proves that with determination, anything is possible. As you will learn, although I began life herding cows and fixing tractors, I made my way to the stars, ultimately overseeing all NASA Civilian Space Science, including Astrophysics (the Hubble and James Webb Space telescopes), Planetary Science, Earth Science, and Heliophysics.

By sharing my journey—the highs and the lows—I hope to show that it's okay to face doubts and setbacks. These moments are part of the process, and they don't define you; how you respond to them does. Resilience and determination have been essential in helping me overcome challenges and grow, both personally and professionally. When the path forward seems uncertain, I want to inspire you to persevere. You have the power to shape your own path to success and fulfillment, no matter how intimidating it may seem.

Don't let anyone's doubts or criticisms hold you back from reaching your full potential—because you are more capable than you think.

CHAPTER 1:
THE EARLY YEARS

Childhood

Growing up on a small dairy farm in western Pennsylvania was quite an experience. Rising before the crack of dawn to help with milking cows, feeding animals, and mending fences instilled in me the value of hard work and responsibility from an early age. I learned how to drive a tractor before I could even reach the pedals and spent countless hours playing in the barn and fields rather than being glued to screens like so many kids today.

Family Farm photo of the early 1970s

Every morning and evening, it was our job to bring the cows in from the fields to be milked. We would also feed them grain in the barn and place silage or fodder beside the lane for them to eat. Someone would drive the tractor with a wagon while someone else used a pitchfork to spread it. Most of the time, eager to be milked, the cows were waiting at the gate, but now and then, they would decide to hang out at the far end of the field. Instead of trudging through the grass to round them up, I would hop on my Honda 350 motorcycle—it seemed like a smart move at the time.

One time, as I was cruising through the tall grass, I completely missed seeing a huge hole in the ground and hit it dead center. I went flying over the handlebars, landing on the grass while my motorcycle tumbled toward me. The engine was still running, almost as if it were laughing at me. I managed to get back on the bike, my ankle throbbing, and rode home, dreading the task of telling my dad what had happened. Of course, he thought I was just trying to get out of work.

I had gotten that motorcycle when I was 16; I saw an ad for a Honda 350 motorcycle in the local newspaper and contacted the seller. The motorcycle was dirty, with oil on the chassis. The drive chain was dry, the engine sputtered, and the exhaust pipes were coated with an oily residue, making it seem that it had not been properly cared for. I didn't pay much money for the motorcycle since it was in poor condition and needed some serious love. I scrubbed off the grease, tightened the drive chain, and made the muffler shine like new. After the test ride, I parked it in the garage. That's when my little brother, who was probably around 5 or 6, saw the shiny area where the tailpipe leaves the engine and decided to give it a kiss. Lesson learned for my brother—always check if the bike is still hot before getting too close.

Working on a farm was no joke—there were so many routine chores to be done every single day! Feeding and milking the cows before the sun was even up was just the beginning. From sunup to sundown, there was always something that needed doing—mucking out the stables, fixing fences, planting crops for the upcoming harvest, and so much more.

We enjoyed simple pleasures like going to a local ice cream parlor with my dad after a long day working in the fields. The ice cream cones were a special treat and didn't happen very often. During the summer months, we spent our lives on the farm in relative isolation from the rest of society, except on weekends when we attended church. The days were long and hot, but there was a certain satisfaction and a feeling of accomplishment in seeing the instant fruits of our labor. At the end of the day, we were physically tired but emotionally charged.

I spent most of my time alone, just listening to the sounds of nature: the crickets chirping, leaves rustling, and the occasional moo from our livestock. It was a peaceful way of life, filled with hard work and simple joys like watching the sunset over the fields. Sure, I missed having company sometimes, but there was something special about feeling so connected to the land and all its changes throughout the year.

Growing up in a low-income farm family as the middle child of eight siblings sure had its challenges. Not only did we have to deal with limited money, but I always felt like I was stuck in the middle—not the oldest or youngest, not the loudest or quietest. It was tough trying to stand out and make a name for myself when I was constantly competing for attention among so many siblings. And the physical demands of farm life left barely any time for fun. My older brother suffered from severe allergies, so he was unable to help with harvesting hay, which also increased the workload.

I never had any allies in the family, being the middle child. One memory that stands out to me is when my family took a trip to Somerset, PA, about 30 miles from our home, so my parents could attend to some paperwork at the courthouse. My siblings and I stayed in the car for about 30 minutes while our parents were inside. I was probably 7 or 8 at the time, and, of course, my siblings couldn't resist picking on me, so they decided to make up a spooky story just to scare me.

They noticed smoke coming from the courthouse chimney and concocted a tale about them burning naughty children. They even told me I had a brother and sister named Orfaorfa and Jerybery. They claimed the smoke was from my siblings being burned for misbehaving and that our parents were inside dealing with my behavior for a possible burning, too. Siblings can be so mean sometimes!

As my parents approached the car, my siblings warned me not to spill the beans about our conversation or else they'd hurt me. Oh, the things siblings do to mess with you!

Despite all the struggles, growing up in that environment taught me valuable lessons about the importance of hard work, perseverance, and the significance of family. As the middle child, I often felt overlooked compared to my older and younger siblings. That's why I turned to our family dog for comfort. Whether we were playing fetch or just chilling in the yard, my furry buddy always had my back when I needed a break from the chaos of being in the middle.

I even used to dream about having a German Shepherd farm of my own one day. It's unbelievable how much joy and comfort a simple animal can bring into your life during tough times.

However, not all dogs were friendly. I remember a few times when our neighbor's German Shepherd would come to our farm and act aggressively toward me, so I had to climb up a tree until the dog left. It was definitely more exciting than running back to the house. Have you ever read the story "Peter and the Wolf"? When I was a young boy, I was really impressed by stories like that one. In the story, a young boy climbs a tree and lassos a wolf with a rope when it comes around. So, I had a rope stashed in the tree, and when the dog tried to get at me, I climbed up and tried to lasso it. I wasn't as successful as the boy in the story, but it was fun to give it a shot.

Growing up, my wardrobe was mostly filled with hand-me-downs from my older brothers or clothes that my mom had sewn for me with love, of course. We were a family on a tight budget, so buying new clothes was never really an option for us. At first, I was a bit embarrassed about wearing secondhand or homemade clothes, but as I grew older, I began to appreciate the charm and personal touch that these pieces added to my style. Even though I didn't have the latest trends or fancy brand names, I learned to embrace and take pride in my unique wardrobe, which helped me express my individuality in a way that money couldn't buy.

As a curious kid, I used to watch and question things that didn't quite add up to me. Our house was a big farmhouse we shared with my grandparents. The kitchen had an old house vibe with a makeshift towel hanger—basically just nails in the wall. One day, I found myself fixated on one of the nails, wondering what would happen if the atoms that made up the wall suddenly separated. Would the nail just fall out? And if it did, would the atoms in the floor also break apart, causing me to plummet to the ground? I pondered these thoughts for about a half hour, amazed that everything was holding together and I didn't have to worry about crashing through the floor.

One of my favorite childhood pastimes was playing in the sandbox, which served as my escape from reality and a portal into a world of endless possibilities. I loved spending hours building imaginary structures and creating my own little world. It was like being a mini architect or engineer, letting my imagination run wild. I would carefully design and build houses, castles, roads, and even whole cities out of sand and toys. Sometimes, I'd add water to create rivers or moats or use sticks and leaves for added details. I felt incredibly proud of my creations, as if they were real works of art. I was also surprised when I came to the sandbox the next day to see that the cats had also enjoyed it and left their deposits. Those carefree moments were among the happiest memories of my childhood, which I'll always hold dear.

We were a large family, but we always had enough food. My mother basically made everything from scratch, even pizza, which helped save money—and it also tasted better. She would typically make a meat-and-potatoes-type full meal; however, one of my favorite memories was when my mom would make a large pot of hot chocolate and fresh cinnamon rolls as a meal in itself. To this day, when I'm feeling under the weather, hot chocolate and cinnamon rolls are my go-to comfort foods. But I also have a soft spot for traditional meat and potato meals that remind me of home-cooked goodness, warming me up from the inside out.

As a child, I learned the Bible verse: "In the beginning, God Created the Heavens and the Earth" (Genesis 1:1), but I did not understand the boundaries of the "heavens" and thought the heavens must be beyond what I was seeing. And while the "heavens" part can be a little confusing as a child, as I matured, I realized that the "heavens" refers to the sky above us, the stars, planets, and galaxies. So, when it says God created the heavens and the Earth, it's just saying He made everything in existence from scratch. It is like when you start a new project and lay down the foundation—that's God setting things in motion with this verse.

I used to spend so many summer nights staring up at the sky, wondering what was beyond what I could see. I'd sneak out of the house, lay on the grass with a blanket, and get lost in the sea of twinkling stars above me. It was a moment of pure wonder and curiosity, thinking about all the possibilities beyond my sight. The quiet of the night made it even more magical, making me feel small yet connected to something bigger than myself. As I looked at the Moon, I wondered if I would be able to see where both Neil Armstrong and Buzz Aldrin had walked.

I'll never forget that one summer night when I decided to count the stars in the sky. I quickly realized it was a lost cause—there were just too many twinkling lights up there for me to keep track of. Instead of getting frustrated, I laid back on my blanket and let myself get lost in the beauty of the endless night sky. The sheer vastness of it all was overwhelming but in a good way. I felt so at peace just staring up at those twinkling stars, letting their light wash over me like a gentle wave. It was a moment of pure clarity and contentment, a reminder that sometimes it's best to just be in the moment and appreciate the beauty of the world around us without overthinking it.

As a child, I was always fascinated by rockets, the Apollo Moon program, and the incredible achievements of astronauts like Neil Armstrong and Buzz Aldrin. Their moonwalks captured my imagination and left a lasting impression on me at a young age. Growing up, our family did not have a television, so we went to our neighbor's house to watch these brave astronauts take those historic steps on the lunar surface. I remember lying on the floor in front of the TV, as I had when staring up at the big, vast sky, feeling a sense of awe and wonder. It was these same feelings that led me to try to build my own makeshift rocket in the backyard.

When I was old enough, I started working as an auction runner to earn some extra cash. The auctioneer usually held auctions on the first Thursday evening and last Saturday of the month. My job was to carry items to the buyers and handle the money exchange. I always got there early to check out what was in those boxes before the auction started. It was like a little sneak peek, and getting a good look at everything up for grabs helped me figure out which boxes I wanted to bid on. Sometimes, I'd find real gems, like antique bottles or a toolset. I ended up saving enough money to buy some really cool model rockets.

Building and launching rockets has been a blast! It was like reliving some of those summer nights looking at the stars and watching Neil and Buzz take their first steps on the Moon. One of the rockets that I bought was the Mercury Redstone rocket, which was used to help test concepts that led to the build-up of the successful Apollo program. Building model rockets was, in my own way, mirroring the Apollo mission. The whole process, from piecing together all the parts to watching my creation soar through the air, was exciting and fulfilling. It was definitely worth every penny I saved up for it!

One of the smaller model rockets

Imagine this: you've got this amazing set of rockets that are just itching to go higher and faster. How do you make that happen? It's simple—just slap on some bigger engines and watch it take off! But here's the catch—if you go too big with those engines, you might end up burning up your parachute on the way up or losing the rocket. It's all about finding that perfect balance between power and control to ensure a successful launch and recovery.

The model rocket engine propellants were solid motors—no moving parts. The rocket models always came with recommended engine sizes ranging from A to D size motors. I had an inquisitive mind and liked to experiment, so I modified the rocket motor compartments to accommodate rockets that the manufacturer specified should use B motors. However, I used D motors instead. I assumed I needed to add additional parachute protection but only guessed at how much to add.

I once added significantly larger engines to a 3-engine rocket, but I didn't properly calculate the amount of parachute protection needed for the bigger engines. The launch was spectacular, with flames shooting out the bottom of the rocket as it soared into the sky. However, when the parachute deployed, all I saw was a melted ball of plastic instead of a functioning parachute. The rocket made it back to the ground safely but at an accelerated speed that partially damaged the rocket's body. Another time, I added a larger engine to a small single-stage pencil-type rocket and watched it shoot into the sky. I didn't adjust the parachute size or add a hole in the center of the parachute to compensate for the higher altitude, so the rocket ended up drifting a quarter mile into the woods, never to be recovered. These experiences taught me valuable lessons about the importance of proper margins in space systems. Adding larger engines to the rockets wasn't a terrible idea, but it definitely had its risks. During my time at NASA, I learned that space systems also require careful design and adequate design margins to prevent failures. Unlike my model rockets, the failure of NASA systems could result in the loss of important scientific information or even cause serious injuries.

Have you ever tried something new, and it totally backfired? Like when you modify a bicycle to go faster and end up with the chain jumping gears? Yeah, it's frustrating. Well, that's exactly how I felt when my brilliant idea to ignite dud rocket engines by setting them on fire went terribly wrong.

I was fed up with these useless rocket engines just taking up space and not doing what they were supposed to do. So, I thought, why not try igniting them up with a small fire? It seemed like a good plan at the time, but it was a disaster waiting to happen.

I set up a fire in the middle of a large gravel area between the garage and barn, using a cardboard oil container to create a mix of gas and oil as fuel. Yes, the same type of container we used for car oil. Certainly not the smartest move.

When I tried to light the rocket engines, things went downhill fast. I thought it would be cool to see them ignite, but boy, was I wrong. Lesson learned: Never mess with dud rocket engines. As soon as the first engine lit, flames and sparks shot out, sending it flying over 400 yards into a nearby field. The second engine did the same, but this time, it was heading straight for the barn, where the doors were wide open, exposing dry hay. I almost started a huge fire, putting myself and my little brother at risk. Fortunately, the lit engine missed the barn.

I quickly realized I needed to stop this madness. I kicked the oil can with the mixed fuel in a desperate attempt to put out the flames, but it ended up landing on my younger brother's jacket and setting him on fire. In a panic, he started to run toward the house to jump in the basement water trough. Our house had a pipe from a nearby spring that filled a cement trough, approximately four by eight feet in size and two feet deep. I tackled him to prevent him from running into the house and causing even more trouble for me. It wasn't the smartest move, but it worked—the fire was out. I guess you could say I performed the "stop-drop-and-roll" without thinking about it. Sometimes, you just get lucky!

My parents didn't find out about the chaos until much later. Lesson learned the hard way—never play with fire, especially when it involves rocket engines.

There have been moments in my life when things got so crazy that I genuinely questioned how I managed to make it through, or I looked back and thought, "How am I still kickin'?" It's like when you're standing on the edge of a cliff or going down a roller coaster—you experience this rush of adrenaline mixed with a hint of fear that makes you stop and think, "Wow, life is wild." Perhaps it's all part of the unpredictable journey we call life—filled with ups and downs, twists and turns.

All year long, I looked forward to attending the County Fair, which came around every fall. I would spend hours walking around the fairgrounds, exploring all the different exhibits, and indulging in delicious fair food, such as cotton candy and funnel cakes. One of my favorite events was watching the demolition derby, where cars would crash into each other until only one was left standing. The excitement in the air was palpable as we cheered on our favorite drivers.

Another highlight for me was the car racing on the dirt track—it was thrilling to witness the speed and skill of the drivers as they zoomed around corners, slid sideways, and raced toward the finish line.

Another highlight was the Joey Chitwood show, where he would not only perform a 180-degree maneuver driving at high speed going forward but then send the car 180 degrees and continue going the same direction backward but also drive the car on two side wheels.

After watching all those exciting adventures, a young child might feel inspired to recreate some of their favorite moments at home. Whether it's pretending to be a race car driver or a stunt driver, the possibilities for imaginative play are endless.

I recall a time when I took a tractor to a back field far from the house's view. I pretended I was racing, using the field as my track. The field had ruts, slopes, and flat areas, so I had a blast driving around my makeshift track, getting more aggressive with each lap. The adventure came to a halt when the tractor tipped on two wheels during a turn, but luckily, it didn't flip over. My parents had no idea about my little escapade until years later. I thought I was being sneaky, but looking back, it probably wasn't the smartest idea. At least I survived and now it's just a funny story to reminisce about.

That tractor experience was no joke! The high center of gravity had me feeling like I was always on the edge of disaster. So, I decided to switch things up and try our family's car instead. I found a flatter field that, unfortunately, was also visible from the house and went for it. What a thrill! The low-slung body made me feel like I was flying on the track. The adrenaline rush was amazing as I zoomed around corners and pushed the limits of the field. It was a stark contrast to the nerve-wracking moments in the tractor, but it felt great to let loose and have fun behind the wheel of my race car. The thrill of driving the car in my mock racetrack was short-lived when my dad confronted me about the tracks in the field.

One important lesson we can learn from this experience is that young people are really impressionable and pay close attention to what others do and say. Whether it's teachers, parents, or friends, kids are always absorbing information like sponges. Adults must keep this in mind when interacting with young people and setting a good example for them. By showing kindness, respect, and positive behavior, we can help shape the attitudes and actions of the next generation in a positive way. Remember, our words and actions can have a lasting impact on young minds, so let's try our best to be good role models and offer guidance that will help them succeed in life.

My time on the farm was a real eye-opener. Taking care of animals and crops taught me to put their needs before my own, which really drove home the importance of being responsible for my actions. It was more than just a job—it was a crash course in building character.

Working on the farm also taught me the value of discipline, integrity, and responsibility. Waking up early every day to tend to the fields or animals showed me the importance of sticking to my commitments and giving my all to my work. It wasn't always easy, but I learned to push through and keep going even when things got tough. And working with nature taught me that honesty and transparency are key when dealing with plants and animals.

School Experiences — The Early Years

In the first grade, I had a genius idea to make extra cash by writing and selling riddles to my classmates. I charged a whopping 2 pennies per riddle, thinking I was onto something big. I spent hours devising clever and humorous riddles to sell during recess, and I even created a menu with different categories, including animals, food, and jokes. Surprisingly, my little side hustle actually took off, and soon enough, some of the kids in my class were buying riddles from me. Looking back on it now, it's quite amusing to think about how seriously I took making those two pennies per riddle. But you've got to start somewhere.

Growing up, I never truly appreciated the value of education, and, as I mentioned earlier, I didn't put much effort into my schoolwork. Honestly, I found it all pretty boring and pointless at times. This mindset led me to underachieve and not reach my full academic potential. Looking back, I can see now how crucial education is for setting a person up for success in the future. Without a solid educational foundation, it's hard to excel in any field or pursue any career path. It's never too late to change your attitude toward learning, though.

I'll never forget my third-grade teacher, Mrs. Johnson (not her real name), who always emphasized the importance of voting in our society. She made sure to explain to us that it was a way for us to have a voice in decisions that affected our lives. However, she also stressed that voting was a private matter and something we should think carefully about before sharing with others. She believed that everyone had the right to their own opinions and should be able to express them freely without fear of judgment or backlash from others. This lesson has stuck with me throughout my life, as I continue to value the act of voting while respecting the privacy of others when it comes to their political belief.

As I mentioned earlier, I was really struggling with math in fourth grade—failing miserably, if you will. It wasn't until many years later that I realized how dangerously close I had come to being left back due to my abysmally low math grade. But my mom tried her best to help me. We would sit at the kitchen table with my math homework spread out in front of us. She patiently went over problem after problem, trying to explain things in a way that made sense to me. I have to give her credit because she really did help me improve over time. It was still a struggle, but having my mom there by my side meant everything to me. As an aside, my mother only had an eighth-grade education, so I wonder how much of a challenge this was for her to help me. Even though math isn't my strong suit, I will always appreciate her dedication and support during that tough time.

In the fifth grade, it was more about hanging out with friends and building relationships than focusing on academics. We still had homework and tests, but it felt like the year when everyone cared more about gossiping during recess or who they were going to sit with at lunch. It was a time of forming cliques and navigating social dynamics within our peer group. Looking back, I see how the social aspect of fifth grade played a huge role in shaping my friendships and experiences.

I also understand that my fifth-grade teacher recognized leadership qualities within me even before I was aware of them myself. I remember how he would often assign me special tasks, like leading group projects or writing a class skit, which made me feel both excited and nervous at the same time. He would offer words of encouragement and guidance, gently pushing me out of my comfort zone to help me develop my skills. It was not until years later that I understood his intentions and appreciated the impact he had on shaping my confidence and ability to take charge.

Reflecting on those moments now, I realize how important it is for educators to identify and nurture potential leaders early on in a child's development, just as my fifth-grade teacher did for me.

CHAPTER 2:
EXTRACURRICULAR ACTIVITIES

Music

In seventh grade, I discovered my love for music and decided to join the school band. I didn't know much about instruments or reading music, so I chose the E-flat horn and later the French horn, following my band director's advice. It was tough at first, trying to make a decent sound instead of just squeaks and squawks. However, as time passed, and with daily practice of the music lessons, I became proficient with both instruments.

Me in 7th grade marching with my E-Flat horn

The E-flat horn's bright, bold sound really added something special to our performances, while the French horn's warm tones brought a touch of elegance. Being in the school band helped me improve my musical skills and make friends with other music lovers. And I was selected to play the French horn in the All-County Band in the tenth through twelfth grades and the State FFA Band in the twelfth grade. After so many years of playing, it felt amazing to be recognized for all my hard work and dedication.

The audition process was no joke, but all those hours of practicing really paid off. Being able to perform alongside some of the best musicians in the state was a truly surreal experience. Plus, getting to travel around and showcase my skills at various events was just the cherry on top.

The experience of attending the All-County Band was not just about the music itself but also about the people I met along the way. Whether you strike up a conversation with someone waiting in line for food or bond with fellow fans during a performance, connecting with others is an integral part of the festival experience.

At the All-County Band practice, I ended up sitting next to this awesome French Horn player. Not only was she a talented musician, but she also had a killer sense of humor. We hit it off right away, cracking jokes in between rehearsals and even during the actual practice. We bonded over our shared love for music, and she even gave me some tips on how to improve my French horn skills. I tended to make the French horn sound brassy rather than mellow. It was great to meet someone so passionate about their work. This unexpected friendship made the whole experience even more enjoyable. Who would have thought that a simple seating arrangement could lead to such a cool connection?

Back then, friendship meant doing things together. So, when my new friend invited me to attend an opera with her family, I was all in. It was a totally new experience for me, but I was excited to try something different and spend time with my friend. And it certainly was quite an experience! The music was beautiful, the costumes were extravagant, and the entire atmosphere of the place was truly elegant. Sitting next to my friend's family, sharing that special moment made me feel all grown up and sophisticated.

After the All-County Band practice and concert ended, it was disappointing that we all lost contact with each other. We had spent so much time rehearsing and bonding over our shared love of music, and then suddenly, everyone went their own way. It's interesting how life can change so quickly.

During band practice, I dedicated a lot of time to mastering the French horn. I experimented with placing the mouthpiece on the side of my mouth instead of the center, switching back and forth from side to side until I could control the tone effectively from either side. I also practiced coordinating different beats simultaneously with my hands and legs, maintaining four distinct rhythms to challenge my mind and improve my multitasking skills, enabling me to play two parts of the music score. Learning to carry different beats with each foot and both hands while thinking of a different beat did not happen instantly; it required significant practice and guidance, which the band instructor was more than happy to provide.

I had the idea to try playing two horns at the same time since I had already mastered playing with either side of my mouth. Surprisingly, I was able to play a simple song simultaneously on both horns. It was a fun and rewarding challenge that pushed me to try new things just for the sake of pushing my limits.

Playing the French horn in the concert band was exciting, but it wasn't really beneficial for the marching band. So, I played the E-flat horn and trumpet in the school marching band and joined the school jazz band playing the trombone. What an awesome experience for me! I absolutely loved being able to showcase my musical talents while also being part of something bigger than myself. The marching band performances were always full of energy and so much fun—we all moved in sync and played our hearts out.

Switching over to the jazz band for those smooth, soulful tunes was just as thrilling. Improvising and grooving with the rest of the band was always a highlight for me. I remember the first jazz band concert where the band director started the song and then walked off the stage to let the jazz band continue on our own; what a thrill! It also gave me the opportunity to explore various music styles and challenge myself as a musician. Overall, being part of both bands really added some extra excitement to my high school years, which were otherwise a troubling experience.

My parents strongly encouraged me to join both the school chorus and the community choral society, believing I possessed a remarkable bass voice. Under their persuasion, I decided to participate. I dedicated myself to the chorus and pursued vocal lessons to enhance my skills. Additionally, my father formed a quartet, which included himself as a tenor, one brother as lead, and another brother as a baritone, while I sang bass. Surprisingly, this experience turned out to be quite enjoyable. Through their persistent encouragement and my diligent efforts, I developed a genuine appreciation for performing and singing in front of an audience.

This journey underscored that sometimes, fulfilling our parents' wishes can lead to unexpectedly rewarding experiences.

Sports

Growing up in PA, I was an avid fan of the Pittsburgh Steelers, the Pittsburgh Pirates, and the Pittsburgh Penguins. Our school could not afford football equipment and offered only soccer, baseball, and basketball as sports options. My favorite soccer player was Pele, so I wanted to mimic him. My older relative was an exceptional soccer goalie, so in my second year of playing soccer, I played the goalie position.

However, balancing practice schedules with farm work was no easy task. Some days, I had to wake up early to do chores on the farm before heading to school, then practice or play games, and finally, head home to do more chores. The discipline I learned from excelling in sports while still helping out on the farm taught me some important life lessons that I still carry with me today.

When I didn't have my driver's license, I would jog home from soccer practice so I could finish my chores on time. The 5-mile jog through the hills was tough at first, but it soon became a routine that I actually came to enjoy. There was something freeing about feeling the breeze on my face and knowing I was getting a workout in while being productive. Sure, there were days when I wished I could just hop in a car and drive home without breaking a sweat, but looking back, those jogs were some of the most peaceful moments of my teenage years. And getting my license after all that hard work made me appreciate having a car even more!

My older brother played high school basketball, so it seemed like a natural fit that I would, too. We set up a space in our barn with a basketball hoop for practice. I tried to balance my love for basketball with my responsibilities on the family farm. But as both demands increased, it became nearly impossible to handle them each effectively. Evening practices and games started cutting into the time I needed to finish my farm tasks. It wasn't just about feeling tired or overwhelmed; it was about being torn between two worlds that meant a lot to me. Eventually, my dad insisted that I quit playing basketball. It was a tough decision, but I knew my commitment to the farm had to come first.

The disappointment of having to quit playing basketball was further frustrating because my parents never attended any of my soccer or basketball games. I put in so much time and effort practicing and giving it my all on the field, and it would have been great to see them cheering me on from the sidelines. It made me feel like they didn't care about what mattered to me. Seeing other kids' parents there supporting them during games just made me feel even more hurt that mine weren't there for me. It definitely affected my motivation and made me wonder if they even cared about my interests or accomplishments.

Motorcycles

When I was 16, I upgraded from a Honda 350 to a 1972 Honda 750 motorcycle. My neighbor had just gotten a Kawasaki 900, which was known as the fastest bike on the quarter mile at the time. Naturally, I couldn't resist trying to make my bike faster. I decided to drill out the fuel jets in the carburetor with a 1/6-inch drill bit. When I took it for a spin, it sputtered at higher RPMs, not exactly the result I had hoped for. I realized I may have messed up the fuel-to-air ratio, so I removed the air filter to allow more air into the combustion chamber. Surprisingly, my Honda 750 now felt comparable to the Kawasaki 900. We never officially raced our bikes, but mine definitely held its own.

These additional extracurricular activities, combined with the leisure time spent on the farm, provided a unique childhood experience. However, not everything proceeded smoothly as I got older.

CHAPTER 3:
HIGH SCHOOL
THE TROUBLED YEARS

In 7th grade, my English teacher had a major issue with how I pronounced my name. She kept insisting that I should be saying it differently, but I knew she was way off. My name is spelled Geoffrey, and it is pronounced the same as "Jeffrey," but she insisted on calling me Goffrey. She said that Geoffrey Chaucer also pronounced his name as Goffery. I talked to my dad about it, and he even went to the school board, but the problem persisted all year. It was frustrating to deal with on a daily basis.

The teacher had a rule that you wouldn't fail her class if you turned in all the homework, regardless of test scores. I saw that as a challenge, so I turned in all the homework but purposely wrote wrong answers on the tests. I ended up with a D- in English class; the teacher kept her word, so I passed. Looking back, I probably could have tried a bit harder for a better grade, but oh well.

This situation with this teacher really got to me and affected my attitude for the rest of high school. Some of my other teachers noticed my negativity. I didn't realize I was bringing everyone down with my bad attitude.

It's important to remember how our behavior can impact those around us, especially when teachers are there to support us.

As I said earlier, one of my teachers once told me I would never amount to anything and would just be a "dumb farmer." Despite not really caring about education or growing up on a farm, what gave a teacher the right to make assumptions about my potential like that? Just because I come from a farming background doesn't mean I can't achieve great things. We all have the power to defy expectations and prove people wrong. So, if you ever face a negative comment like that, use it as motivation to show them what you're really capable of. You might just surprise everyone, including yourself.

In another situation, my math class had a test to determine the top math student in the school. I was told I couldn't take the test because there weren't enough score sheets, and I wouldn't need the score for college anyway since I wasn't planning to attend college.

Then, my guidance counselor discouraged me from taking college prep classes because he didn't think I was college-bound.

Everyone has a unique path and potential, so don't let anyone's doubts hold you back. If you want to go to college, go for it. Taking those college prep classes can still benefit you in other ways, such as preparing you for future opportunities and challenging you academically.

After taking the advice of my guidance counselor, I decided to switch things up and focus on something other than the usual academic classes. I chose not to take the SAT because I was told that college wasn't in the cards for me. It just didn't seem worth it to spend all that time and money on a test that wouldn't really benefit me in the long run.

My FFA jacket

Instead, I joined the Future Farmers of America (FFA) program and was ultimately chosen as the school's FFA president. Who would have thought that joining the FFA would lead me to become the school FFA president and County vice president? But there I was! Being a part of FFA allowed me to learn about agriculture, leadership, and teamwork. It gave me a sense of purpose and belonging in my school community, and being selected as president was both an honor and a responsibility.

Let me tell you about my time in the school Vocational Agriculture (Vo Ag) class—it was a blast! I got to roll up my sleeves and dive headfirst into rebuilding engines. From tearing them down to putting them back together, I learned a great deal about how these machines work and how to troubleshoot any problems that might arise. The hands-on experience was priceless—getting greasy and digging into the nitty-gritty of these engines really gave me a sense of accomplishment. Plus, working alongside my classmates made it all the more fun, instilling in me a sense of teamwork and camaraderie. I won't mention the time we took the Vo Ag teacher's pickup for a winter drive and got it stuck in a snow-filled ditch. Imagine explaining why we were late for the next class.

I was totally surprised when my classmates picked me to be their Senior Class President. I never thought I'd be given such a huge responsibility. It was a real honor to know that my peers trusted me enough to lead them through our last year of high school, especially since I wasn't exactly on an academic track. My classmates voted me as the most talented and the most likely to succeed, which was quite unexpected. One of my classmates even asked me to relinquish the'

Most Likely to Succeed' title so he could have it instead, which would help him with his college plans.

We had several important decisions to make, including choosing a theme for prom, arranging and securing approval from the school board for our senior class trip, and determining how we wanted to leave our mark on the school before graduation. Looking back on it all, I believe my classmates recognized leadership qualities in me that I hadn't even realized I possessed.

I had somewhat of a flair for adventure, so I invited President Jimmy Carter and Rosalynn to my senior graduation. I wrote a short letter that said, "From one President to another, I am pleased to invite both you and Rosalynn to my high school graduation." Two weeks later, I received this postcard.

Postcard I received from President Carter

As the Senior Class President, I was proud to have the responsibility of representing our class by giving the senior class speech. I had the opportunity to address all my peers and share some heartfelt words of wisdom and encouragement with them before we all headed off on our separate paths. I spent weeks preparing for it, making sure every word was just right and that my message was truly meaningful. My speech was based on the tree analogy, with its base as a foundation and branches as opportunities. When the big day arrived, I felt a mix of nerves and excitement as I stood up there in front of everyone, but once I started speaking from the heart, those nerves melted away.

Have you ever attended a ceremony where they pass on some form of leadership or authority? It's called the ceremonial passing of the mantle, and essentially, it involves symbolically transferring power or responsibility from one person to another. I've actually participated in this ceremony before and was ceremoniously passing the mantle to the junior class president. I have to admit, I tried to mess with the pin holding up the mantle on the junior class president's shoulder. It was one of those moments when I was feeling mischievous, especially since I would be presenting the mantle to an ex-girlfriend and just wanted to see what would happen if I accidentally stuck the pin. Who hasn't had the urge to mess with things they're not supposed to? Anyway, my attempt failed miserably, and nothing happened. Looking back, it was childish, but it created a memory that still makes me chuckle.

I graduated from high school in the lower half of the class, probably closer to the lower third. I definitely wasn't at the top, that's for sure. My less-than-stellar academic experience was reflected in my class rankings. However, graduating is an achievement, regardless of where you land in the rankings.

Since then, I've worked hard to better myself and my skills. Although I wasn't a straight-A student back then, I'm proud of how much I've grown since.

What a crazy turn of events! Who would have thought that years after leaving high school, I'd be asked to come back to the school and give the commencement speech at graduation? When they reached out and invited me to be the speaker, I was so humbled and proud to share some wisdom with the graduating class. And it was a great chance for me to prove all those doubters wrong—you know, the ones who said I wouldn't amount to anything. So, of course, I said yes right away.

CHAPTER 4:
THE BEGINNING

First job

After graduating from high school, I landed a job as a maintenance worker at a local meatpacking plant. Let me tell you, it's not the most glamorous job in the world, but someone has to do it! My daily tasks involved fixing and maintaining the machinery used in the meat processing process to ensure everything ran smoothly and efficiently. Whether it was repairing processing equipment or troubleshooting electrical systems, I was the go-to guy for any maintenance issues that popped up. It can get pretty messy and smelly, but I learned to embrace the unique challenges that come with working in this environment. In reality, this job was just an extension of what I learned growing up on the farm.

Married life

I got married to Lauretta on my 20th birthday! Having our wedding on my birthday was a strategic move to ensure I wouldn't forget our anniversary, as the birthday cards served as effective reminders. There were occasions when I almost forgot our anniversary, but the timely arrival of the birthday cards prevented any embarrassment.

It may sound crazy, but when you find the one, you just know. We were young and carefree, embarking on this journey as husband and wife—and it's been one wild ride. We've had our fair share of ups and downs, like any couple, but at the end of the day, I couldn't imagine doing life with anyone else but Lauretta by my side. In fact, July 2025 marks our 45th wedding anniversary.

After graduating from high school, the curiosity I had as a kid for seeing how things worked didn't just disappear. We bought a new car, a Ford Escort, the first model year they were produced in the 1980s. The garage informed us that the improved fuel efficiency was related to the shape of the pistons but wouldn't disclose the details. I couldn't just let that slide. So, when the car reached around 40,000 miles, I decided to remove the engine head to inspect the pistons and see for myself what made them so special. It was a simple change from other pistons, and it would have been easier for the garage just to tell me the answer, but I had to find out for myself.

While trying to understand the new engine design, I also looked for other ways to improve fuel efficiency and came across an ad for a water injector. This thing sprayed water directly into the carburetor based on engine speed. The idea was that the water would turn to steam and give more compression power, ultimately boosting fuel efficiency. At least, that was the theory. So, I bought the water injector kit and modified it so I could read the water injector spray speed. What could possibly go wrong? In the winter, we added dry line antifreeze to get rid of any water in the gas tank, but then I was spraying water straight into the carburetor.

One winter day, we were driving back from LaVale, MD, and were about 20 miles from home when the engine started acting up and struggled to climb a small freeway hill. Could it be water freezing in the carburetor? You bet it was. I turned off the water injector, let the engine's heat melt the ice, and we were back on track. But I wasn't ready to throw in the towel just yet. I decided to add alcohol to the water injector tank to prevent icing. Not knowing the right mixture necessary, the car still had periodic freezing issues, so I just dumped pure alcohol into the tank, replacing all the water. What was I hoping to achieve? Better fuel economy, of course, and lower operating costs. One issue was that the cost of alcohol outweighed any potential fuel-saving benefits I was hoping to see. You never know until you try, though. Luckily, my experiments didn't end up harming the car.

Our son entered the world

A year after we tied the knot, our son Eugene entered the world with a sweet demeanor and bright eyes that immediately stole our hearts.

Eugene's middle name pays tribute to Lauretta's late brother Morris, who tragically passed away in a car accident. The birthing experience was quite unique compared to many others our age. Before we got married, Lauretta worked for her dad in his medical office located in a rural Amish community in Ohio. The Amish community was hesitant to have their babies delivered at the local hospital, so Lauretta's dad and another doctor set up a special house equipped with essential medical tools for delivering babies. The house was always spotless but lacked electricity. Lauretta had witnessed many childbirths with her dad and saw the compassionate care provided in that home setting and wanted Eugene to be brought into the world in the same way. I was a bit unsure at first, but after researching the success rate of births at that home versus the local hospital, I had no objections.

Eugene made his grand entrance around 7:15 pm. I vividly remember the doctor using what looked like a miner's light during the delivery. The Amish host also cooked a mouthwatering 4-course meal for us. It's funny how certain details stick with you.

Eugene brought so much joy and light into our lives with his infectious smile and charming personality. Watching him grow and hit milestones like his first steps and first words has been a delightful journey for us as parents. We feel incredibly fortunate to have Eugene in our lives, filling our family with love and laughter every single day.

I had left the meatpacking plant for a sales job. Unfortunately, after our son was born, my job went downhill, and we ran out of money. It was tough—first dealing with the chaos of being new parents and then dealing with financial stress. We had to make some tough choices about our spending and budgeting, which meant saying goodbye to some luxuries and finding ways to save money wherever possible. It was a nerve-wracking time as we tried to figure out when or if things would get better. We were falling behind on rent and car payments, with unexpected expenses constantly arising. We tried to keep up, but it just wasn't happening. It felt like we were always playing catch-up, struggling to make ends meet. We had to decide which bills to pay first and had some tough talks about where we could cut back.

One night, we enjoyed a dinner of boxed macaroni, the kind where you empty the box into a pan, add water, and heat until the cheese melts around the macaroni, accompanied by a refreshing glass of water. It may not have been the fanciest meal, but it was all we had left in the house except for baby food, and we couldn't afford to buy more. Talk about stress!

Later that night, there was a knock on the door. I opened it to find someone who I was not fond of and his young son standing there. He said they had extra food and wanted to share it with us because they noticed we were struggling. How did he know we were out of food? I tried to keep our situation to ourselves. Some might call it divine timing, but I think there was a bigger lesson at play. I had to decide whether to let my pride get in the way or accept his offer. It was a tough choice, but I swallowed my pride and accepted the food.

Lesson #1: His act of kindness warmed my heart and reminded me of the power of compassion in our community. Instead of holding onto negative feelings, focusing on the good in people can make a big difference. The man's genuine smile and the boy's eager eyes made me grateful for such thoughtful neighbors. It was a simple gesture of generosity that had a lasting impact on me.

Soon, I landed a job as a sheet metalist and welder for a high-voltage test equipment manufacturing company. It was a big change for me, but I was excited to learn new skills and challenge myself. The work was tough and physically demanding, but I quickly grew to love the hands-on aspect of the job. I mastered the art of welding intricate pieces of metal together with precision, creating structures to support high-voltage test equipment. Seeing my hard work come to life in a tangible way was incredibly satisfying, knowing it played a crucial role in keeping things running smoothly. Overall, it was a rewarding experience that taught me valuable skills and expanded my knowledge in the field.

Just when I thought our finances were back on track, life threw me another curveball. One day, while chatting with my coworker during our break, we delved into deep conversations about our passions, work projects, and general interests. Lloyd, who usually kept opinions to himself, opened up about his time as a radio technician in Vietnam. He rarely talked about it, but when he did, you could see the conflict in his eyes. He described the intensity and chaos of his experiences, highlighting the importance of staying calm and focused under fire to maintain communication lines. Despite downplaying his role, it was clear that what he did made a significant impact.

He noticed my interest in electronics and suggested that I take some evening courses at the local community college to improve my skills. I never expected it, but it's amazing how a random chat led me down an entirely new path in my career. I had never really thought about taking night classes before. I told Lloyd about my high school days and joked that I didn't think I was smart enough for college, a seed of self-doubt that had been implanted by teachers in high school.

Out of nowhere, he said, "Let's enroll in college together." Although I again shared my high school struggles, he insisted he'd take the class with me if I agreed to attend. We decided to enroll in a Digital Electronics 101 course. And there we were, sitting in class next to each other, trying to figure out circuitry and programming. It was actually pretty fun having someone to lean on and bounce ideas off of. We both aced that course together.

It suddenly hit me like a ton of bricks that maybe I was not the person those high school teachers believed I would be—just a farmer with no other capabilities and certainly not college material. It was a wake-up call that I needed to step up and start taking more initiative in my personal and professional life. I later realized that Lloyd could have taught the course but wanted to provide me with support and encouragement.

Lesson #2: This experience taught me to look for others who needed that extra boost or figurative hand-holding, providing encouragement and mentoring, and served as a guiding principle throughout my career.

The experience at community college showed me that I was capable of college, and I couldn't pass up the opportunity to further my education. At the same time, my wife and I decided it was time for a change. After much thought, we agreed that I should leave my job and go back to school for a 4-year degree in engineering. It was a tough decision, but we knew it was the right move for my career and our future. Having my wife's support made me feel even more confident in my decision. It was going to be a challenging journey, but we were excited to see where this new chapter would take us.

When I started applying to engineering colleges, the impact of my poor track record in high school came to the forefront. I thought my passion for tech would be enough to get me in, but I was wrong. My high school grades and lack of an SAT score made college acceptance difficult. I knew then that my dream of studying engineering might not come true, which was a tough pill to swallow. However, instead of giving up, which I could have easily done, I researched ways to improve my application to demonstrate to colleges that I was serious about pursuing engineering.

I was excited when I found out that Capital College, a highly prestigious private college in Laurel, Maryland, had accepted me into its Telecommunication Engineering program. The only catch was that I had to prove myself during my first year by not repeating my high school performance. I wasn't eligible for scholarships in my freshman year, so I took out student loans and found a part-time job to help cover expenses.

I tried applying for a position in the maintenance department at the college, but unfortunately, I was rejected. I believed that my experience on the farm and with maintenance made me the perfect fit for the job. Instead, I ended up working part-time in the school library, which turned out to be a blessing in disguise. I was grateful to be in a cozy environment during extreme weather rather than out in the elements.

My schedule was pretty hectic, balancing being a husband and father and full-time classes with part-time work, but I enjoyed the challenge. Working in the library allowed me to be surrounded by books and help my fellow students find the resources they needed for their assignments. It was tough, but it taught me valuable time management skills and how to prioritize tasks effectively. Overall, even though it was busy, working in the school library really enhanced my college experience.

I was worried about getting through my freshman year without the benefit of college prep classes in high school. Other classmates talked about their prep classes, including calculus, and thought the college Calculus I class would be a breeze. I started to panic. I signed up for Calculus I, took extensive notes, and studied diligently. Despite my lack of high school preparatory classes, I earned an A, while some students who took preparatory classes struggled and even dropped out of college due to poor grades. Not only was I better at math than my fourth-grade failure had prophesied, but it also proved that one can succeed with determination.

Lauretta and I were both working hard to make ends meet, but we also had to swallow the bitter pill of paying for childcare on an already stretched budget. We did not have any relatives or family close to us, so we needed to find someone to watch our son during the day. A mutual friend agreed to watch him for a reduced rate, but this was still a financial challenge. When our friend became pregnant, she decided to stop watching Eugene, so we needed to find an alternate, more expensive person to provide daycare. It wasn't easy seeing a chunk of our paycheck disappear every month, but knowing that our little guy was in good hands made it all worthwhile. Watching him learn and grow in daycare was truly priceless. We knew, at the end of the day, that we were making the best choice for our family, even if it meant sacrificing some financial freedom.

I always made sure to set aside time each week to spend with Eugene. We loved going to ham (computer-related and not meat) fests to hunt for new gadgets and relics to buy. Every Saturday, we rode our bikes to Dunkin' Donuts for a sweet treat. I discovered Eugene's passion for card collecting, so we started going to card shows and autograph events together, providing quality father and son time. We collected over 200,000 cards and numerous autographs from baseball and football players. This shared hobby created a special bond between us that remains strong to this day.

In my junior year of college, I decided to switch majors from telecommunications to engineering. It was not an easy choice, but after careful consideration and consultation with academic advisors, I realized that my true passion lay in electronics. I was drawn to the hands-on nature of electronic engineering and the endless possibilities for innovation and problem-solving that this field offers. Plus, I felt that electronic engineering offered a more diverse range of career opportunities that aligned better with my long-term goals.

My telecommunication classes were a valuable asset in being selected for a Co-Op at ARINC in Annapolis, MD, where I gained hands-on experience testing ground-to-air communications for commercial aircraft. During my time there, I learned programming, troubleshooting, and how to work effectively with leadership.

After my experience at ARINC, I continued my Co-Op at Litton Systems in College Park, MD. This opportunity enabled me to further my education while gaining hands-on experience in military systems, space projects, and commercial ventures. It was truly exciting to be involved in such diverse industries!

This experience taught me the importance of effective communication and strategic planning in high-pressure environments. I was even hired as a full-time employee at Litton Systems while finishing my senior year of college.

Working on space projects was a highlight for me, as I got to witness firsthand how cutting-edge technologies are developed and implemented for exploring the unknown. Overall, my Co-Op experiences opened my eyes to the endless possibilities within each industry and solidified my passion for pursuing a career that makes a real impact on the world.

I graduated college with a 4-year degree and Magna Cum Laude status. It's remarkable to think that I was once told I wasn't smart enough for college, and now I've achieved this milestone. Walking across that stage to receive my diploma with honors was an incredibly rewarding feeling.

Lesson #3: Not applying myself in high school resulted in higher stress levels during college. Actions have consequences!

Have you ever experienced something that completely changed your perspective on life? These kinds of experiences have a way of shaking us to our core. They force us to face our fears and insecurities head-on, prompting us to reevaluate everything we thought we knew about ourselves and the world around us.

In 1998, my parent's farm in Meyersdale, PA, was hit by not just one but two tornadoes in a devastating turn of events. The damage caused by these natural disasters was significant and left a lasting impact on our family. Trees were uprooted, and buildings were destroyed within minutes. It was a chaotic and surreal experience seeing the aftermath of such powerful forces of nature firsthand. It was a challenging time, but it also brought us closer together as we worked through the ramifications of the destruction. The memory of that fateful day still lingers, serving as a reminder of the unpredictable and destructive power of Mother Nature.

My family, neighbors, and friends were all a wonderful source of support. However, when things became even more overwhelming, the Red Cross and the Salvation Army stepped in, providing much-needed support. The Red Cross provided not only food but also emotional support during moments of crisis. Their volunteers were always there with a helping hand and willing to lend a listening ear. Similarly, the Salvation Army offered assistance with clothing, household items, food, and financial aid. Both organizations not only helped meet our immediate needs but also provided comfort and a sense of security during some of the darkest moments in our lives.

The devastation caused by the tornado went far beyond just the physical destruction of buildings and infrastructure—it also resulted in the heartbreaking loss of two family members. On that fateful day, while I was working at Litton Systems in College Park, Maryland, I received a call about my parents' farm being struck by a tornado. Without hesitation, I jumped in the car and headed from College Park to Meyersdale, Pennsylvania. It was a long drive, but family always comes first.

As I started the four-hour journey, my phone rang with more devastating news. My sister's husband and daughter had tragically passed away from carbon monoxide poisoning caused by a generator running in a confined space in the basement. Despite numerous public warnings not to run gas-powered items in a confined space, my brother-in-law disregarded the warnings. He placed a generator in the basement to keep the refrigerators running. I suppose he thought there was sufficient ventilation in the basement to clear the carbon monoxide from the generator. It was a heartbreaking and surreal experience trying to come to terms with the sudden loss of two beloved family members. The drive felt never-ending as I grappled with the reality of a world without them in it. I began reminiscing about lighter subjects, like the time a tornado hit my uncle's farm when I was a young boy. I heard from my parents that a tomato destroyed my uncle's farm; at least, that's what I thought I heard. I couldn't wrap my head around how a tomato could cause so much destruction. As I was driving, the movie *Fried Green Tomatoes* came to mind. Looking back, it seems pretty silly, but it helped me stay calm in the moment.

On my way to the farm, I stopped at my sister's house and was the first family member to arrive. The house was still buzzing with medical and police activity. I was advised not to go inside since the bodies were still there. My sister, who worked night shifts at a nursing home, had come home in the morning to the house filled with carbon monoxide and deceased loved ones, was overcome with carbon monoxide. Fortunately, emergency personnel came to the scene in time, rescued her, and rushed her to the hospital before I arrived. In my daze, I couldn't even remember her last name when the police asked me about the family. Stress can have a profound impact on your mind.

Tornado damage from two tornados just days apart

As I arrived at the farm, I couldn't believe the sight that greeted me. Destruction and debris were scattered throughout the fields as I made my way down the quarter-mile lane. The house and barn were in shambles—parts of the house's roof were missing, the barn had been demolished, and neighbors and friends were already starting the clean-up effort. My mind was flooded with memories of growing up in that house, working in the barn, playing in the sandbox that was now covered with fallen trees, and spending time in the garage doing autobody work on several cars. My childhood memories were just that, memories, as the actual items were damaged or destroyed. I looked for my old Honda 750, but it, too, was destroyed in the tornado.

With my sister's sudden loss, I knew I had to step up and help her through the difficult process of planning their funerals. Despite feeling my own overwhelming grief, I put on a brave face and focused on being there for her every step of the way. It was tough to hide my own pain in order to stay strong for her, but knowing how much it meant to her to have someone there to support her during such a devastating time made it all worth it.

This is one instance where I went through a really tough situation that forced me to learn how to hide my emotions in order to be strong for others. At the time, it was all about putting on a brave face and pretending like everything was okay when, in reality, I was struggling inside. It wasn't easy, and it definitely took a toll on me emotionally, but I knew deep down that it was necessary to support my sister when she needed me most.

Lesson #4: Through that experience, I discovered the importance of being there for others, even when you're going through your own pain. While I may have put off grieving until later, it ultimately taught me resilience and the power of perseverance during difficult times.

I completed my BS degree, but getting into NASA would prove to be in a league of its own. It took me 12 years of hard work and perseverance before joining NASA at the Johnson Space Center in Houston, Texas. It turns out my leadership skills still needed some fine-tuning along the way. Lying in the grass, staring into the vastness of the universe, my dream began so many years ago. Although it took longer than expected, I never gave up. As a child at the time, I didn't know it was NASA—I just knew there was an organization exploring the stars, and I wanted to be part of it.

Litton Systems had previously offered me a full-time job as a Reliability Assurance Engineer, so I continued on that path, building the necessary skills. The opportunity to continue growing my career with a reputable company like Litton Systems is something that doesn't come around every day. This job provided me with the necessary skills, including technical problem-solving, management, time management, teamwork, and the importance of paying attention to details, which ultimately led me to NASA.

You never know where a job might take you if you're willing to put in the effort and seize every opportunity that comes your way.

As a Reliability Assurance Engineer (RAE), my primary responsibility was to ensure that all systems were running smoothly. This involved reviewing parts, checking drawings, addressing any issues that arose, and monitoring problems to determine if they were related to a larger issue. Once, I encountered a challenging problem that left me stumped. I tried various tests, but none seemed to work. However, that night, my brain must have been working overtime because I woke up in the morning with the solution. I tested out my theory, and sure enough, it was right.

I still remember being given the role of Qualification Test Director for a Navy aircraft avionics upgrade project. It was a big responsibility but also an exciting opportunity. Essentially, my role was to ensure that all testing procedures and protocols were in place to guarantee the safety and functionality of the new avionics system and to oversee the testing process. I had to coordinate with engineers, technicians, and even the prime contractor brass to get everything sorted out. It was a challenging task, no doubt, but I learned a great deal from it. I remember putting in long hours, troubleshooting issues left and right, and feeling like a total boss when everything finally came together seamlessly, making it a career highlight.

After completing the Navy project, they handed over the reins to me as the Qualification Test Director for a secure communication system. Essentially, my role was to ensure that this new technology was up to standard and ready for deployment. I had to put it through its paces, running various tests and ensuring everything worked as it should. Again, it was a significant responsibility but also a considerable challenge. I had to stay on top of all the details to ensure nothing slipped through the cracks. I'm always up for a challenge, and getting to work on cutting-edge technology like this was definitely worth it. Plus, knowing I played a key role in keeping our communications secure was a pretty sweet feeling.

One day, I was sitting in my boss's office, discussing my job, when he dropped a bombshell on me—he wanted me to be the

Engineering Project Manager (EPM) for the systems used on the LA class submarines and Navy aircraft. These systems were super important for early detection and warning during missions. As EPM, I was responsible for handling all technical aspects and providing final approval on all drawings. I was excited but also a little nervous about the gig, as it meant taking on more responsibility and a chance to showcase my leadership skills.

My boss had faith in me, though, and promised to have my back every step of the way. Although I was feeling a bit overwhelmed at first, I was thrilled that my hard work was being acknowledged. I left his office feeling fired up and ready to crush this new challenge.

Looking back on these experiences, it's clear that the leadership skills I gained in previous roles were preparing me for my time at NASA. From leading group projects to making tough decisions under pressure, I was consistently challenged and pushed out of my comfort zone, allowing me to grow and develop as a leader. Yet there were still more challenges ahead before I was offered the NASA job.

The Chief Operating Officer of Litton called me and asked me to serve as a corporate troubleshooter for several issues with two of the Divisions. My job was to investigate any unusual activities occurring at various company Divisions. It was a bit out of the blue, but I suppose my reputation for fixing things preceded me! They needed someone who could think quickly and handle tough situations, so naturally, they thought of me. I'll admit it's a little nerve-wracking to take on such a big role, but I was ready for the challenge. Plus, there's something thrilling about diving headfirst into a high-pressure situation and coming out on top. On one occasion, they even sent me on a one-way flight, promising a return ticket once I cracked the case.

So, picture this: I was the corporate troubleshooter on one job site, and the troubleshooting efforts kept getting extended. It was like a never-ending cycle of delays. Finally, on the last extension, I just couldn't take it anymore. I had plans with my wife that I kept having to cancel because of this project. So, I asked my boss to call my wife and let her know what was going on. And he actually did it! He called her and explained the whole situation. It was definitely a bit embarrassing having my boss talk to my wife, but desperate times call for desperate measures! In the end, she totally understood, and we were able to reschedule our plans.

To put it into perspective, I was sent to one of the Divisions to determine the cause of failure for a critical item used by the military. The Division was losing significant dollars per week due to the delay and was also at risk of being placed on the poor performance list. I was given three days to identify the cause and propose a solution. On the third day, the Litton COO and the responsible military organization person in charge of the project came to the Division and wanted answers. Sometimes, things just work out. I was able to identify the cause of failure quickly and provided recommended solutions for the brass at the meeting. Fortunately, my recommendations resolved the issues, and the project was back on track.

Under the "other duties as required" category, I took on a variety of additional tasks, such as troubleshooting issues, assisting different teams, and handling unexpected challenges as they arose. I contributed to a Navy-funded study on solder joint failures in surface-mount components, through-hole components, and high-density connectors. We assessed the best equipment for repairing, replacing, and installing these components, and our findings were included in the Navy's best practices documentation.

I found myself in a pivotal position that ultimately led to my opportunity to work for NASA. I was assigned as Task Manager for the manufacturing and delivery of the Space-to-Space Communications system to NASA. Astronauts use this system during their Extra-Vehicular Activities (EVAs) to communicate with the Space Shuttle and the International Space Station. In other words, it's a big deal—this system is crucial for keeping those brave men and women connected while they're doing spacewalks. In an attempt to identify the problem, I made multiple trips to the Johnson Space Center, where I interacted with my NASA counterparts.

SSCS plaque was presented to me by JSC for my contributions to successfully flying the system. The flag was flown on Shuttle Atlantis STS-106.

We were experiencing issues with the SSCS system not performing as expected or not meeting our expectations. It was a bit of a headache trying to pinpoint the root cause of the problem, but we continued working on it collectively. It was frustrating because we were relying heavily on the SSCS system to work smoothly for us, yet we had to contend with these hiccups.

During one of my visits to JSC, I felt that the issue was related to the JSC design rather than our manufacturing and testing process. It was one of those days where nothing seemed to be going right, and frustration was running high. In a moment of exasperation, I turned to the Division Chief and commented that he should hire me to resolve these types of issues. Sure, it may have sounded a bit bold and audacious, but desperate times call for desperate measures, after all. And let's be real here—my skills were clearly needed. Sometimes, a little boldness can go a long way in showcasing your worth and landing that dream job.

The Division Chief commented that I should submit my resume and application paperwork to the NASA Human Resources Department. Within two weeks, I received a nice postcard from NASA stating that they appreciated my interest in NASA but that there were no opportunities available. I felt confident in my abilities, thanks to the foundation laid throughout my career thus far, yet now it seemed my dream was out of reach.

Lesson #5: The lesson is that patience is a virtue when it comes to making decisions. Although I thought I exhibited fortitude in preparing to work for NASA, I was going to have to stretch the limits of my patience if I truly wanted to achieve my dream.

About six months after the exchange, I received a phone call from the Division Chief asking if I was serious about working for NASA at the Johnson Space Center. He said that the Center received authorization to hire three "critical hires" and that he was authorized to offer me one of the positions. The catch is that he was unsure how long the authorization to hire would stay in place and requested that I provide an answer relatively quickly. He said that if I didn't want the job, he needed to offer the position to another candidate before the opportunity was lost.

Accepting the offer would mean relocating from Maryland to Clear Lake, TX, leaving our son in MD as he started his softmore year of college at the University of MD, and leaving friends to make new ones in Texas. Two hours after receiving the phone call from the Division Chief, and believing this was a once-in-a-lifetime opportunity, I accepted the job without even discussing salary or other key details, knowing they would follow.

When opportunity doors open, especially if it's something you have been preparing for, don't hesitate to make your move. Taking calculated risks and capitalizing on opportunities can lead to great success. On the other hand, don't force a move if the timing doesn't feel right. Rushing into things can lead to mistakes and missed opportunities. As someone once said, "Don't wait until all the stop lights are green before starting to cross the city."

Lesson #6: Knowing when to act can make a significant difference in achieving your goals. *Stay patient, stay focused, and seize the moment when it presents itself.*

As you will learn, I decided to apply these key principles throughout my NASA career. This just goes to show how important it is to take stock of what we've learned and apply those lessons to navigate life's twists and turns.

CHAPTER 5:
DREAMS CAN COME TRUE

Having breakfast with our son before we set out to drive to Texas was a truly special moment for us. We sat around the table, savoring each bite of pancakes and bacon as we talked about the road ahead.

As we said our goodbyes and started the long drive to Texas, tears filled our eyes, knowing that we were leaving our son behind. It was a bittersweet moment, filled with a mix of pride in his accomplishments and sadness at the distance between us. We had spent so much time together, watching him grow and mature into the young man he had become. Now, as he embarked on his own journey in Maryland, we couldn't help but feel a twinge of loss in our hearts. Despite the sadness, we knew that this separation was necessary as I started my new chapter with NASA. So, with heavy hearts, we left him behind, knowing that he would always be loved and supported no matter where life took him.

I had spent twenty years preparing for my dream of a career with NASA. From gazing up at the star-filled sky to the education challenges and setbacks to the many jobs, roles, and responsibilities, I knew that I had been preparing for this opportunity my entire life. But deep down, I couldn't help but wonder if I was really prepared and whether all that I had learned and experienced could, in fact, be applied to this next journey.

As we drove to Texas, I recalled some of my most memorable experiences with Eugene. After announcing my departure from Litton, I had three months before I started working at NASA. We often visited the golf driving range, computer shows, and other venues, enjoying our time together. At first, though, three months felt like a long time, but in reality, it passed quickly.

When he was young, Eugene and I often played with a football in the backyard, sometimes filling it with helium to see if it would go further. We tried kicking and throwing the ball, believing the helium really helped the ball fly further. In reality, who knows if it helped.

But this led to another exciting and memorable event. When Eugene was a senior in high school, he and I played the mellophone with the University of Maryland marching band at a World Football League halftime show in Barcelona, Spain—all the weeks of practice leading up to the flight to Spain seemed to pay off. We spent days promoting the World Football League game by marching through the streets of Barcelona—an amazing father-son bonding experience.

As I mentioned earlier, Eugene and I would often go to autograph events for professional athletes, and one event in particular stands out. Johnny Unitas, famed quarterback for the Baltimore Colts, signed Eugene's football. To make the event more memorable, Unitas even signed Eugene's cast, which, unfortunately, he was wearing on his wrist. When it came time to remove the cast, Lauretta ensured the doctor avoided damaging the autograph during the removal process. I preserved the cast and still have it today.

As we continued on the road to our future home, the prospect of using my knowledge and expertise to make a difference in space exploration kept me motivated. Even if the path ahead was uncertain, I knew that the journey itself was invaluable in shaping me into the person I was meant to become.

We pulled into Clear Lake, TX, a few days before I was supposed to begin my new job on July 3, 2000. It was nice to have some extra time to settle in and explore the area before jumping into work. Clear Lake is a charming little town with friendly locals and beautiful scenery all around—especially the sparkling lake that gives the town its name. Lauretta and I spent those first few days checking out local restaurants, strolling along the waterfront, and getting our bearings in our new home away from home. It was a relaxing start to what would soon become a busy work schedule, and I was grateful for the chance to acclimate to my surroundings before diving into my new job adventure.

On July 3, 2000, I officially began my new adventure with NASA at the Johnson Space Center. My big job was to basically pave the way for a whole new Branch named EV5 in the Avionics Systems Division.

My charter was to establish a Flight Hardware Development Branch, later renamed the Parts, Package, and Manufacturing Branch within the Avionics Systems Division. I had to roll up my sleeves and define how it would fit into the rest of the organization. From fleshing out the roles and responsibilities of the team members to determining the goals and objectives of this new Branch, there was a lot on my plate. It was like starting from scratch and building something from the ground up.

One challenge was populating the branch with personnel from other branches. I was an unknown entity and needed to establish trust with the new team. Reflecting on my past management experience, I made a conscious effort to become a servant leader and not a micromanager. This meant treating everyone with respect and providing clear vision and direction while allowing the team the freedom to implement the direction without micromanaging every aspect of their job. This approach was crucial for the team to have buy-in as it maximized the likelihood of success. I pushed the team to excel and meet commitments. I stated that I would provide a safety net if they experienced failures as long as they committed to their best effort.

The branch was required to participate in various aspects of developing human spaceflight hardware, including EEE Part selection, ionizing radiation testing, electronic layout, and system design. We weren't just focusing on one aspect either. And, of course, it wasn't always smooth sailing. There were setbacks and challenges galore, but we never gave up. In many cases, the astronauts were the ones using the items we built, so I tried to have the team attach an astronaut's name to the item to add a sense of both pride and personal responsibility.

It was truly a once-in-a-lifetime experience working with the astronauts who would be using the systems we built. We had the opportunity to see firsthand how our technology would be put to use in space, and it was both thrilling and nerve-wracking at the same time. The astronauts were incredibly knowledgeable and enthusiastic about their work, making collaboration with them all the more rewarding. It was inspiring to know that our work contributed to the success of such incredible individuals who risked their lives to explore the unknown depths of space. Astronauts face many risks with their missions, but they are real people. They paid attention to detail but took the time to explain why they were concerned about different aspects of the design or elements of their mission. Even with their esteemed positions, they always treated me with respect, and in fact, I maintained friendships with several astronauts throughout my career.

As many of these space explorers were my friends, I still remember the sinking feeling in my stomach when I received the phone call on that fateful day in February 2003 informing me that the Columbia Space Shuttle had broken up during reentry. It was the worst day of my life; tears filled my eyes, and I was in disbelief and shock as I watched the news coverage unfold before me. The tragic loss of those brave astronauts aboard the shuttle was devastating, not just for their families and loved ones but for the entire nation as well.

During a moment with a colleague, Bill Culpepper, who was analyzing the last communications from the astronauts in an effort to gain insight into the issues that caused the shuttle to break up, I was reminded of all the risks and dangers involved in space exploration. The images of debris scattered across the fields will forever be ingrained in my memory, serving as a somber reminder of the fragility of human life and our aspirations to reach beyond the stars.

The Agency sought ways to detect shuttle wing damage during ascent for future missions. One option was attaching a LIDAR-type camera to the shuttle arm to inspect the wings in orbit. The systems needed to be developed safely and quickly so that NASA could start flying the shuttle. I visited a Canadian company where coworkers were working on the inspection camera, a crucial step in identifying issues during the launch of future shuttle missions. The urgency to find a solution, coupled with the need to ensure the system worked properly, reminded me of my days as Qualification Test Director at Litton.

I was later asked to investigate an issue with the Simplified Aid for EVA Rescue (SAFER) backpack, which astronauts use to move untethered in space. Although not typically my Branch responsibility, I applied my experience as a corporate troubleshooter at Litton to identify the cause of the failure, determine the corrective action, and verify the fix.

EEE parts can be a bit confusing at first, but essentially, they refer to any electrical, electronic, or electromechanical components used in various devices and systems. This can include components such as resistors, capacitors, transistors, diodes, and integrated circuits. These parts are essential for the functioning of everything from smartphones to cars to satellites. Each part has its own unique purpose and characteristics that contribute to the overall performance of a system. Understanding how these EEE parts work and interact with each other in the space environment is crucial for designing and troubleshooting electronics used in both human and robotic missions.

Ionizing radiation and EEE parts have a pretty interesting relationship. EEE parts can be affected by ionizing radiation in ways that may not be immediately apparent. When these parts are exposed to ionizing radiation, like X-rays or gamma rays, it can cause damage at the atomic level. Ionizing radiation is a form of energy that has enough power to knock electrons from atoms, creating charged particles (or ions) in the process. This damage can lead to degraded performance or even the complete failure of the component. Engineers must take this into consideration when designing systems for use in environments where ionizing radiation is present, such as outer space or nuclear power plants. It's kind of cool how something so tiny can have such a significant impact on our technology! That's why it was crucial to test for radiation effects on electronic hardware intended for space, where it will be exposed to high-energy protons or heavy ions.

We routinely traveled to the University of Indiana to test electronic components and devices using their high-energy proton beams. I remember spending long nights in Indiana testing hardware only to have a squirrel take down the electrical system, shutting down the proton beam. Bill Culpepper said, with a laugh, "This will take a while, so go back to the hotel and get some rest." When Bill said that, he was probably not kidding around. If something goes wrong and the beam needs to be recalibrated or fixed, it can indeed take many hours or even days before everything is up and running smoothly again. And knowing Bill, he's usually pretty laid back and takes things in stride, so his laughter was probably just his way of acknowledging the situation and trying to lighten the mood.

On one occasion, we were testing laptop computers for use on the International Space Station. We had just turned on the proton beam, and the laptop computers immediately stopped working. Imagine what would have happened in orbit if we had not tested the computers before sending them to space. We needed to find a different lot and brand of computer that worked in the ISS radiation environment.

Another Branch responsibility was maintaining the process and standards for the circuit card layout. Essentially, circuit card layout involves designing the physical placement and connections of components on a printed circuit board (PCB) to ensure optimal performance and functionality. It involves determining the size and shape of the PCB, as well as the placement of components such as resistors, capacitors, and integrated circuits, in a manner that minimizes signal interference and maximizes efficiency. Factors such as signal integrity, thermal management, and manufacturability are all taken into consideration when determining the layout of a circuit card. Ultimately, a strategic circuit card layout is crucial for ensuring that electronic devices operate smoothly and reliably.

It wasn't long before the team was working together like a well-oiled machine. It was exciting and rewarding to see all my hard work pay off as we launched this new branch and watched it thrive within our division. I appointed Dave Bevely as my chief engineer and moved his office to my office suite. Dave and I became best friends outside of work, often riding our motorcycles or just working in his garage, repairing motorcycles. It's amazing how something as simple as a shared interest in motorcycles could lead to such a strong bond, but I'm grateful for it every day. Dave wasn't just a friend; he became a kindred spirit who knew me better than most.

The Branch participated in Agency and international Parts discussions, sharing challenges and experiences. I had the pleasure of presenting the EV5 EEE parts approach to the Japanese Space Agency in Tokyo, Japan. This was my first time presenting to a big group with a translator by my side. It was nerve-wracking, to say the least. I had prepared my speech meticulously and practiced in front of the mirror countless times, but having someone translate my words into a different language added a whole new level of pressure. I had to make sure I spoke clearly and at a pace that allowed the translator to keep up. It was challenging but also incredibly rewarding. Seeing the audience nodding along and engaged despite the language barrier made all the nerves worth it. Plus, having a translator there helped bridge the gap between the audience and me in a beautiful way.

While in Japan, I attended a reception the night before the meeting and tried sushi for the first time. As someone who prefers meat and potatoes, this was new for me. With no other options, I gave it a try. Of course, they provided me with training chopsticks since I was not even a novice with them. It wasn't bad, but I still prefer my meat and potatoes.

The trip to Japan taught me a great deal about effective communication across language barriers.

The Branch was finally up and running, which was a big relief. I knew I needed someone to help me out—enter the Deputy Branch Chief. This person was going to be my right-hand person, sharing in all the responsibilities and tasks that came with running a successful branch. From overseeing day-to-day operations to managing staff and addressing any issues that arose, having a solid Deputy Branch Chief was going to make my job a whole lot easier. Plus, it's always nice to have someone to bounce ideas off of and collaborate with when navigating uncharted territory.

When I first began my job search, I knew one of the key factors I wanted to prioritize was someone with a diverse range of opinions. I had a saying, "I am not smart, you are not smart, but collectively we are brilliant." I also sought someone with the potential to grow and develop within the Branch. I hired a deputy, an extremely competent person who met the criteria. I assured them that I would provide mentorship and prepare them to take over the Branch Chief role when I moved on to other opportunities.

Within three years of selecting my Deputy, they were ready to take on the role of Branch Chief, so I accepted the role of Deputy Division Chief (acting on rotation) within the Avionics Systems Division (ASD). As the title suggests, ASD was responsible for Avionics systems within the Johnson Space Center. This was a key stepping stone to transferring to NASA Headquarters in Washington, DC.

CHAPTER 6:
TRANSFERRING TO NASA HQ

O n January 14, 2004, President Bush announced a Vision for Space Exploration to revitalize the United States' space program. NASA established the Constellation Program as the implementing program, set up at NASA Headquarters in Washington, D.C. I transferred from JSC to HQ in DC to help with setting up the program office. The Program Office was later removed from HQ and established at JSC with oversight in the Exploration Systems Mission Directorate at NASA HQ, but I remained at NASA HQ in DC.

As I prepared to transfer to NASA Headquarters in Washington, DC, I reflected on events at JSC and across the country. On 9/11, while sitting on an airplane taxiing for a flight from Houston to LAX, the pilot informed us of a hold. We soon learned about the attacks. Lauretta called me, expecting to leave a message, but was surprised when I answered. I assured her I was okay, though a sense of helplessness set in. Hearing about the plane in Shanksville, PA, reminded me of playing basketball against the Shanksville team in high school. I want to extend a special thank you to the rescue teams who risked their lives that day.

I had the pleasure of supporting various initiatives while at the Johnson Space Center, which paved the way for me to transfer to NASA Headquarters in Washington, DC, to take on a more senior role in these endeavors.

Securing this opportunity was a total game-changer for my career. Moving to NASA Headquarters in Washington, DC, felt like a huge step forward in my professional journey. Being based at HQ opened up numerous doors for me—I was able to connect with key NASA officials and gain valuable insights into how things worked on a larger scale. It wasn't easy getting here, but man, it was worth every bit of effort.

CHAPTER 7:
ANOTHER DEVASTATING EVENT STRUCK THE NASA FAMILY

I was at my office in NASA HQ in Washington, DC, with the NASA channel playing in the background. I was multitasking, surrounded by the buzz of conversations and the sound of keyboards clacking away. The atmosphere was lively yet focused.

One Friday afternoon in 2007, I glanced up at the screen and saw the latest updates on NASA missions and launches. Suddenly, I noticed emergency vehicles outside Building 44 at JSC. Building 44 was where my office used to be when I worked there in Texas. I decided to reach out to one of my old bosses, Pat Pilola, to find out what was happening in Building 44.

I vividly remember the moment Pat broke the news. There had been a shooting in the building; almost everyone was evacuated, and only three people were left inside—Dave Beverly, my administrative assistant, and a support contractor. The contractor believed that Dave was about to have him fired and proceeded to shoot Dave with the administrative assistant in the room. Dave tragically lost his life trying to protect the administrative assistant from the shooter, who ultimately took his own life in my office. It was a heartbreaking and unimaginable tragedy.

Several years earlier, Dave had requested additional help to complete some of his tasks, so we hired a contractor to provide this much-needed support. I couldn't help but wonder if I had missed something during the interview process. Could I have done something to prevent it?

I gave Linda, his widow, some space to grieve after the horrific event, not wanting to overwhelm her. The next morning, I called her to check in and offer my support. Dave and I were not only coworkers but also best friends. It was a good conversation, recalling the motorcycle group's Saturday breakfasts and working on the motorcycles in Dave's garage. I offered to help in any way I could. Little did I know she would actually take me up on it! She asked me to sing the song "Sheltered in the Arms of God" at two memorial services—one at the JSC main auditorium and the other at the church. I agreed, and we ended the call.

However, I had a dilemma—I didn't know the song and wasn't sure if a soundtrack was available. As a bass singer, I was concerned it might be outside my vocal range. I explained the challenge to Lauretta, who immediately started calling Christian bookstores to find the soundtrack. Meanwhile, I packed for my trip from Washington, DC, to Houston, TX. Luckily, she found a soundtrack, so I grabbed my portable CD player and headphones and learned the song on the 3-hour flight to Houston.

Singing at my best friend's memorial service in front of coworkers and on live TV was a challenge I knew I had to overcome. I sat in the green room getting ready for the service at JSC, periodically looking out at the packed auditorium. TV cameras lined the aisles, and I was told the service would be broadcast live on the NASA channel and one of the major networks' internet channels. It was definitely a surreal experience.

Drawing on my experience of hiding my emotions when my sister's husband and daughter passed away, I knew I had to reach deep inside myself for the strength to do it again at this sad event. While I was on stage, memories of my time with Dave came flooding back. I told the audience about the time we went motorcycle riding, and I couldn't keep up with him. I blamed it on my smaller bike, a Honda Pacific Coast 750 cc. So, Dave sold me his Suzuki GSX1100. But even with the faster bike, I still couldn't keep up with him.

When I finally caught up to him, he said with a straight face, "I don't think the issue is the bike." The audience laughed at that memory and later told me they could see Dave's humor shining through.

Me singing at my best friend's memorial service at JSC

As I sang, my hand was shaking, not from nerves but from the flood of emotions I was feeling. The lyrics brought back memories of the struggles we had faced together. Looking into Linda's eyes, I saw a mix of love, pain, and hope reflected back at me. It was a powerful moment that words couldn't describe.

It's amazing how a simple smile or kind word can totally turn someone's day around. But on the other hand, a mean comment or rude gesture can really stick with someone and bring them down. It just goes to show that our actions, no matter how small, can have a big impact on others without us even realizing it.

A year after the memorial service, I was in Salt Lake City, Utah, for a meeting at a convention-type hotel. While I was waiting in line to check in, a lady approached me and asked if I was Mr. Yoder. I said yes, but I was curious about what she wanted. She told me she had seen the memorial service online and was moved by my song. She hoped to meet me one day to say thank you.

CHAPTER 8

NASA
THE SKY'S THE LIMIT

***

Stepping into my role as Division Director for the Directorate Integration Office (DIO) at NASA Headquarters marked the beginning of a transformative chapter in my career—one filled with innovation, collaboration, and moments I'll never forget. From crafting lunar strategies on napkins during lunch meetings to late-night calls with astronauts like Buzz Aldrin, every experience became a thread in the fabric of a story that's deeply personal to me. Reflecting on these moments now, I'm reminded of how pivotal it is for educators and mentors to recognize leadership potential early on—just as my 5th-grade teacher did for me. That early encouragement helped shape a path that would eventually lead me to represent NASA in international discussions, brief NASA advisory councils, and contribute to some of the most ambitious space exploration efforts of our time. These experiences are more than career highlights—they are memories I hold close to my heart and will always cherish.

Directorate Integration Office

My first key role at HQ was as Division Director for the Directorate Integration Office (DIO). The Directorate managed requirements for the Constellation Program and planned future Lunar missions.

The Constellation program involved multiple centers, posing a challenge for the IT department to develop an integrated system for capturing requirements, design documentation, and risk management.

During a lunch meeting with the IT manager, we discussed challenges and solutions. We brainstormed and documented key points of one solution on a napkin, signing it as an informal agreement.

A major highlight of my time at the DIO office was co-leading—and later leading—the Lunar Architecture Team, which developed strategies for lunar surface operations. Our team evaluated various aspects necessary for lunar habitation, including mobility, habitability, power generation, protection, and optimal traversal routes.

One evening, I received an unexpected call from Astronaut Buzz Aldrin, the second person to walk on the Moon. He had reviewed our draft recommendations and wanted to share his insights. We engaged in an extensive conversation, discussing various options and drawing parallels between his experiences on the lunar surface and our study.

Additionally, I had the responsibility of briefing the NASA Advisory Council on the progress of our study. It was particularly noteworthy to present our findings to the Honorable Harrison "Jack" Schmitt, the only geologist who has walked on the lunar surface as part of Apollo 17.

Our architecture team meetings were sometimes entertaining and challenging, but everyone aimed for a successful outcome. On one occasion, when I pointed out a flaw in a concept, Pat used several Pepsi cans, rearranged the desk, and proceeded to illustrate a potential update to the architectural pattern. Although it may not have been the most conventional method, it proved effective.

The architecture wasn't solely a NASA effort; it also included strategies from other space agencies. During a trip to the Netherlands with a coworker, as we were leaving the airport, I noticed a flash beside the road but didn't really pay attention to it since I wasn't driving. About a mile later, we observed another flash, which I learned was a speed camera. My coworker, Rich, received two speeding tickets upon our return to the States. This experience is why I prefer not to drive in foreign countries, as I'd likely end up driving on the wrong side of the road or missing some other nuance about their driving laws.

The Agency was leading an international effort to develop a Framework for Explanation as part of the Lunar Architecture activity. On one occasion, a meeting was held in Germany with representatives from 13 international space agencies, discussing concepts and potential contributions to the overall architecture. As the senior NASA official at the meeting, I led these discussions. That evening, while sitting in the hotel, I reflected on the journey that led me to this position. I was humbled and in awe that a small-town farm boy could achieve these opportunities despite being told in high school that I would only be a farmer.

On another occasion, I attended a meeting in Florida where we presented our architectural concepts. During a break, several TV stations, including the Disney Channel, conducted interviews with me. In preparation for future interviews and press conferences, I underwent intensive media training, which included being videotaped while trained media personnel asked me questions and set up potential traps. So, by this point, media interviews had become second nature. Our LAT studies had drawn to a conclusion and we were concentrating on the Constellation requirements verification when the Constellation Division Director (CSD) announced he was leaving the Agency. I was directed to take the role of CSD Director.

Constellation Systems Division Director

I had worked closely with the CSD director on Constellation activities as DIO Director, so this transition was a natural fit.

The Constellation Program was developed to facilitate the return of humans to the lunar surface for extended missions. This initiative included learning how to efficiently transport humans and supplies to and from the Moon, extracting resources from lunar regolith to aid in constructing habitats and enhancing mobility on the lunar surface. A critical objective of the Constellation mission was to develop the capability to live on another celestial body, specifically the Moon, while remaining close enough to Earth to safely return astronauts in the event of an emergency. The scientific elements of the mission were formulated using a comprehensive global strategy involving multiple international space agencies.

Plaque presented to me by the ESMD AA
for my contributions to the Constellation Program.

I was responsible for working with the Constellation Program Manager to provide HQ direction and support. Constellation was a complex, integrated program that required multiple design reviews to ensure the system came together as planned, including requirements reviews, progress reviews, and other key milestone reviews. During this process, I had the opportunity to work with test pilot and astronaut John Young, who commanded Apollo 16 and was the ninth person to walk on the Moon. It was particularly rewarding to speak with individuals who had firsthand experience of what the Moon looked like from their perspective.

An important aspect of my role involved overseeing the Commercial Crew Development Program (CCDEV), which provided opportunities for the commercial sector to enhance its space flight and support capabilities. As Director, I was the source selection official for the CCDEV 1 contract. After extensive reviews and collaboration with the evaluation team, I made the selection decisions. On the day I informed the NASA Administrator and Deputy Administrator of my decision, I was required by the legal team to sign the selection documentation and hand it over before the briefing. This procedure was designed to prevent any appearance of external influence on my selection decisions.

Reflecting on my responsibility to oversee both the NASA Constellation program and the CCDEV projects, it sometimes felt like managing two different objectives. However, this approach has led to the development and implementation of multiple methods for space travel.

As the DIO Director, I co-chaired the Lunar Architecture Team (LAT) LAT-1 and chaired the LAT-2 Lunar architecture studies, where we examined potential activities on the Lunar surface. Later, as Constellation Division Director, we reviewed Mars Design Reference Mission (DRM) studies to develop a reference architecture for missions to Mars.

The trip to Mars involves various constraints, including the greater distance, which impacts environmental conditions, landing options, and the required supplies. I participated in the Mars Design Reference Architecture 5 and was one of four individuals who approved its publication. In discussions with colleagues, we debated the number of launches required to preposition supplies on Mars before human arrival, taking into account orbital dynamics and specific launch windows. Our thought was that the prepositioned items needed to be on the Martian surface and verified as being safe before launching humans on the 9-month journey to Mars. The total number of items and the associated weight required to send humans to Mars and sustain them while waiting for the launch window to open for a return to Earth was challenging. Pre-launching supplies and verifying they were safely on the Martian surface was the best way to keep the astronauts safe on the Martian surface once they arrived there. DRA 5 served as the reference architecture.

Geoffrey Yoder

The human exploration of Mars would be a complex undertaking. It is an enterprise that would confirm the potential for humans to leave our home planet and make our way deep outward into the cosmos. Though just a small step on a cosmic scale, it would be a significant one for humans, because it would require leaving Earth on a long mission with very limited return capability. The strategy and implementation concepts that are described in this report should not be viewed as constituting a formal plan for the human exploration of Mars. Instead, this report provides a vision of a potential approach for human Mars exploration that is based on best estimates of what we know. This is the latest in a series of Mars reference missions that are used by NASA to provide a common framework for future planning of systems concepts, technology development, and operational testing. In addition, this architecture description provides a reference for integration between multiple agency efforts including Mars robotic missions, research that is conducted on the International Space Station, as well as future lunar exploration missions and systems. The strategy outlined in this report was developed from the 2007 Mars architecture study. The Mars Architecture Working Group (MAWG) was comprised of agency-wide representatives from the Exploration Systems Mission Directorate (ESMD), Science Mission Directorate (SMD), Aeronautics Research Mission Directorate (ARMD), and Space Operations Mission Directorate (SOMD). A Joint Steering Group of NASA Agency senior leadership was established in January 2007 to providing oversight, guidance, and ultimately concurrence of recommendations made by the MAWG. This strategy will be updated and revised as we learn more about Mars as well as the systems and technologies that are necessary to conduct human exploration missions beyond low-Earth orbit.

Geoffrey Yoder
Director, Constellation Systems Division
Exploration Systems Mission Directorate

Director, Mars Exploration Program
Science Mission Directorate

Thomas H. Irvine
Deputy Associate Administrator
Aeronautics Research Mission Directorate

David F. Radzanowski
Deputy Associate Administrator
Space Operations Mission Directorate

Signature Page for the Design Reference Mission 5
Mars Architecture Steering Group

The Constellation Program was facing both financial and schedule challenges, and it became clear to me that with a change in Administration, it would likely not continue in its current form. The Science Mission Deputy Associate Administrator approached me to see if I was interested in transferring to the Science Mission

88

Directorate (SMD) as the Astrophysics Division Deputy Director. I accepted the challenge and transferred to SMD.

Astrophysics Division Deputy Director

The primary objectives of the Astrophysics Division are to elucidate the universe's workings, comprehend our origins, and address the fundamental question: Are we alone? It may sound simple, but it's actually quite complex. Managing a diverse range of projects and creating new programs is no easy task. Take, for example, the Hubble Space Telescope (HST). It's a crucial observatory that helps us explore the universe and beyond.

Building on its success, the James Webb Space Telescope was developed to take this exploration even further, with capabilities 100 times greater than Hubble. The Hubble Space Telescope was upgraded to include Infrared (IR) capabilities, allowing the telescope to peer through cosmic gases. The HST upgrade was performed in space 350 miles above the Earth.

Although I was an engineer and not a scientist, these scientific missions were extremely exciting, with breakthroughs around every corner.

Missions like the Kepler space telescope search for planets that are deemed to be within the habitable zone. The definition of a "habitable zone" is the distance from a star at which liquid water could exist on the surfaces of orbiting planets. Habitable zones are also known as Goldilocks' zones, where conditions might be just right—neither too hot nor too cold—for life.

When I took over the role of Deputy APD, only two bodies were identified as meeting the habitable zone requirements. Over 2000 bodies now meet this criteria. Another exciting aspect of the APD was the numerous studies, research projects, and technology development efforts that contributed to shaping new missions.

As an engineer in the world of science, I had to bridge the gap between engineering and science, understanding every aspect of our missions. I spent at least 20% of my time collaborating with APD scientists, engineers, and program executives to ensure we covered all bases. I believe in inclusivity among our staff and the acceptance of ownership across all personnel, so I tasked the audiovisual group with creating a movie that showcases our key missions, including interviews and meetings with scientists, engineers, program executives, and administrative staff. I wanted everyone to feel like they're part of something big and to be proud to share our missions with their friends and loved ones.

Unfortunately, not everything always goes as planned. One of my responsibilities as APD Deputy was to oversee the Balloon project managed by the Wallops Flight Facility in Virginia. The NASA Balloon Flight program used two types of balloons: zero-pressure and super-pressure balloons. Super-pressure balloons (SPB) are capable of lifting a scientific payload weighing up to 8,000 pounds. When fully inflated, the SPB volume is 92 times greater than that of a typical blimp. Put another way, when fully inflated, an entire football stadium could fit inside the balloon. The primary objective of the NASA Balloon Program is to provide high-altitude (above 100,000 feet) balloon platforms for scientific and technological investigations.

According to the NASA Balloon website, these investigations lead to fundamental scientific discoveries that enhance our understanding of the Earth, the solar system, and the universe. Scientific balloons also serve as a platform for testing new instruments and spacecraft technologies that align with the Science Mission Directorate Strategic Plan.

1st photo: Credit NASA Balloon Program Office. Loading helium into the balloon

2nd photo: Credit NASA Balloon Program Office, expands to the size of a football stadium

The Balloon Program is a cost-effective means of gathering important scientific data that can be utilized for future, more complex missions. For example, Nobel Laureate Dr. John Matters started his concept evaluation as a balloon project to validate some of his theories. The successful validation led to additional mission concept validation and, ultimately, a satellite mission (COBE), where the findings led to the NOBEL Laureate award.

While launching the balloons was usually routine, the April 2010 launch in Alice Springs, Australia, was anything but. Strong surface winds caused the balloon to drift away from the launch area, ultimately colliding with a parked car before gaining altitude. This incident led to the grounding of the balloon program until the cause could be identified and addressed.

I visited the Wallops facility in Virginia to meet with the balloon team. As I sat in a room filled with balloon management and operators, a sense of nervousness seemed to hang in the air. The silence was palpable, and I could tell they were bracing themselves for a potential lecture on carelessness or incompetence.

My goal was to figure out what was going on without pointing fingers. So, I started the meeting by asking a series of questions, but the answers I received were somewhat vague. I could tell everyone was feeling a bit nervous, so I had them all hold out their hands and then slap one hand with the other. I told them to consider their hand slapped and that we needed to focus on solving the issue, not placing blame.

After that, the team seemed to relax, and we had a really good discussion about the problem and how to move forward. I later discussed the experience with Balloon Program Manager Dave Perce, who said the team was expecting to be reprimanded but was relieved that I was more interested in understanding the issues than scolding them.

Within a year, the failure investigation team had figured out what went wrong, made some changes, and the balloon program was back up and running.

Our first cruise with Celtic Thunder group

As time went on, I realized I needed a break from work—a vacation, to be exact. My wife is absolutely obsessed with the Celtic Thunder Irish group, so when I saw they were hosting a themed cruise, we decided to give it a try. It was our first cruise, and we had no idea what to expect. The performers were amazing, and the port stops were thrilling, and that's when our love for cruising began. I was wearing my APD SOFIA shirt on the cruise.

SOFIA was a modified 747SP aircraft equipped with a 2.5m (8.2 ft) telescope, which enabled observations across the globe from an altitude of 45,000 feet. This altitude is significant as it allows the airplane to fly above 99% of the water vapor in the Earth's atmosphere. The telescope's observation stations were integrated into the aircraft, allowing teachers to participate in observations and inspire their students. It was always thrilling to board the aircraft and witness the remarkable fusion of aviation and scientific capabilities. The scientists' enthusiasm was truly infectious.

Additionally, conversing with the pilots was a pleasure, given that they were operating an aircraft with a specially adapted fuselage. Naturally, all modifications had been rigorously verified to ensure safety.

I was presented this SOFIA plaque by the SOFIA Program Office for my involvement with SOFIA.

Heartbreak struck our family once again when my older brother, who was two years older than me, was diagnosed with glioblastoma, a type of brain cancer that is usually a death sentence. Unfortunately, my brother was no exception. The doctors predicted that he had about 17 months left to live, and sadly, he passed away around that time. It was a harsh reminder that life is unpredictable.

As my brother's health declined, I made the weekly three-hour drive to his house to be by his side. Sometimes, all I could do was listen, talk, or hold his hand. Memories of our time together flooded my mind as I thought about the inevitable. A special memory was when I collected sports memorabilia to donate to a local school for their fundraising auction. I remember walking into the school after traveling from Virginia to support the auction. My brother acted as if I was some type of sports scout or legend and played it up throughout that evening.

My coworkers noticed my weekly trips from Virginia to Pennsylvania to be with my brother, and they mentioned how I showed the same care and concern for them.

> People indeed pay attention to your actions, whether they're good or bad. It's important to show compassion and support to those who need it most.

When I was appointed acting APD Division Director, I wasn't sure what to expect from the science community since the previous director was a scientist, and I was an engineer. However, I presented my approach to the National Academy of Science and the NASA

Astrophysics Advisory Committee and received their full endorsement. Months later, several Astrophysics scientists approached me and appreciated my fairness in not favoring one discipline over another, attributing it to my engineering background.

One time, I flew to Paris to meet my European Space Agency counterpart for the first time in person. We had spoken on the phone many times before but never face-to-face. During the flight, I thought about our previous conversations and the topic we were going to discuss in Paris. We decided to have dinner the night before our meeting. And boy, did I make an impression.

We were seated at a nice restaurant in a cozy booth with white tablecloths. The conversations were all casual at that point. As I reached for some bread, I accidentally knocked over a glass of red wine, spilling it all over the table, my white shirt, and the booth wall. I calmly remarked, "Sometimes my introductions are rather unusual." We both laughed, and that moment marked the beginning of a great partnership.

The Agency was making changes to how they carried out independent assessments, and I was asked to lead the newly formed Office of Evaluation (OOE). I gathered the APD team together to share the news that I would be transitioning to my new role in the OOE. The APD team is like a close-knit family, so it was tough for me to break the news and say goodbye.

Office of Evaluation Director (OOE)

After a change in administration, NASA began evaluating various aspects of the Agency. One area under scrutiny was how the Agency conducted independent assessments. I was asked to lead the newly formed Office of Evaluation, which was responsible for implementing all independent assessments across the Agency and reporting directly to the NASA Administrator. This was a change for me, as I was previously involved in program execution, and now I was responsible for conducting independent assessments. This turned out to be a benefit because, at times, I believe some of the traditional independent assessments were largely "check the box" exercises and not necessarily beneficial to the Agency.

I focused on developing value-added assessments for the Agency rather than the checklist approach of previous groups. I swiftly established the key elements of the office, including hiring a Deputy OOE Director. The team updated policies and procedures for both technical and programmatic assessments based on the new guidance I established. Just as I was getting into the swing of things and starting to enjoy my role, I was then asked to take on the position of Program Director for the James Webb Space Telescope (JWST).

JWST Program Director

I was approached by NASA's Associate Administrator, Robert Lightfoot, about taking on the position of JWST Program Director. However, I politely declined because I still had unfinished business with the OOE job. When I was asked a second time, I declined again. Then, NASA Administrator General Charles Bolden Jr. called me into his office and strongly suggested that I consider the JWST Program Director position. I mentioned my unfinished tasks with OOE and expressed concerns about some of the underlying issues brewing with the JWST. General Bolden responded by saying, "That's why I am directing you to take the position of JWST Program Director. Nobody else was telling me there may be problems." I accepted the challenge graciously.

So, what exactly is the James Webb Space Telescope? It's a cryogenic, 6.5-meter-diameter space telescope with a unique design driven by its science requirements, a passively cooled cryogenic design, and the need to stow the observatory for launch.

I quickly got up to speed on the program and the international partnership aspects of the telescope. I appreciated the progress and leadership of the project management team and focused on building a partnership rather than an adversarial relationship as Program Director/Project Director. Our common goal was to successfully launch and operate the telescope and make groundbreaking scientific discoveries.

We met weekly at the project office at the Goddard Space Flight Center to review progress and address any potential issues.

One of my notable experiences was traveling to Paris to meet with my counterpart from ESA for the JWST project. The meeting required an in-person discussion. I departed from Dulles Airport in the evening on a night flight to Paris, arriving early in the morning. After checking into the hotel, I traveled to the ESA facility for a four-hour meeting before catching a return flight to Dulles Airport later that afternoon. Reflecting on this experience, it was a highly successful trip from a program perspective, although physically demanding. It shouldn't be that difficult, but sometimes meetings are that important, requiring a face-to-face discussion.

I had the pleasure of working alongside Dr. Eric Smith as my deputy and Program Scientist. Eric and I would regularly visit Capitol Hill to discuss the progress and issues surrounding JWST. It was crucial to keep our stakeholders fully informed about the status of JWST. These visits also provided us with the opportunity to engage with the community, including schools.

One memorable experience was when I traveled to Pittsburgh to speak with over 300 middle school students in their auditorium about JWST. I emphasized that everyone can be a part of these exciting missions, even if they are not scientists or engineers. Dr. Eric Smith and Astronaut Leland Melvin joined us via the Internet to share their unique perspectives on science and education.

I will never forget the children's excitement when Leland talked about his time in space and showed a video of himself putting an M&M inside a water bubble, explaining the science behind it. The students were fascinated by how science, math, engineering, and technology all played a crucial role in JWST. They were particularly intrigued by how the telescope needed to be folded up for launch and then unfolded for operations, comparing it to their Transformer toys.

Overall, it was gratifying to witness the students' enthusiasm for space exploration and the significant role that the James Webb Space Telescope plays in advancing our understanding of the universe. Several weeks after the event, I received a copy of an article about the event from a local newspaper, highlighting my discussions with the students. It is encouraging when others recognize and appreciate the efforts to educate our youth.

One of my favorite JWST images is the towering pillars of creation, a vast expanse of sculpted gas and dust located approximately 6,500 light-years away in the Eagle Nebula. They have been a well-known landmark in the Milky Way since they were first observed by the Hubble Space Telescope in 1995. The shimmering portrait of these iconic structures captured by the JWST could provide scientists with new insights into the process of star formation and how stars influence the space surrounding them.

Eagle Nebula, Credits: NASA, ESA, CSA, STScI; Joseph DePasquale (STScI), Anton M. Koekemoer (STScI), Alyssa Pagan (STScI).

But I was about to have another directed career change.

Deputy Associate Administrator for Programs (DAAP) for the Science Mission Directorate

The incumbent DAAP retired, and I was assigned the role of DAAP. While the DAAP's responsibilities overlapped with the DAA, the DAAP focused on the technical oversight of the Science Mission

Directorate missions and supported the DAA with programmatic oversight, with the primary focus being technical. My boss and I informed the Hill staff that I would be leaving the JWST Program Director position to take on the role of DAAP. The teleconference was met with concern as a staffer felt it was not a responsible decision to remove me from the JWST PD role. He stated that I have been open and honest with JWST's progress and issues. We assured the staffer that as DAAP, I would still oversee the JWST along with many other missions, totaling 114 missions. The staffer reluctantly accepted this explanation.

I had a policy of visiting successful bidders within three months of the award to show support for the mission before any issues arose, helping build a team concept focused on success. This approach contrasted with the selection process, where a thorough assessment of all aspects of the proposals was required to ensure we selected the best projects.

I recall a visit to one of the institutions to meet with the newly selected team for an informal lunch. During the visit, an elder team member remarked, "Thank you for visiting us. I have been here for 20 years, and you are the first senior person to visit us for a casual chat." This comment was significant but also indicated a need for more senior interaction with the team.

One policy I implemented was an open-door policy, encouraging staff to thoroughly consider their questions before approaching me. This facilitated richer discussions and typically led to better outcomes. I often emphasized the importance of collaborative decision-making.

Occasionally, I would argue a position contrary to my own beliefs to explore different options.

Maintaining open and honest communication was a priority. I recall a conversation at the LADEE launch at Wallops Flight Facility. I expressed frustration about cost issues with another mission, stating that they could not accurately predict funding needs even after funds were expended. This straightforward dialogue between myself and the GSFC Center Director was effective in quickly getting to the heart of issues and resolving them rather than allowing them to persist. My reassignment to elevated positions continued when the SMD Deputy Associate Administrator left the Agency.

Deputy Associate Administrator (DAA), NASA Science Mission Directorate

I was asked to step into the role of DAA after the previous DAA left the Agency. It seemed like a trend that I was being reassigned whenever someone left, which could give the impression that I couldn't hold down a job. In reality, these were directed promotions showing the confidence my management had in my abilities.

As the DAA, I was responsible for providing executive leadership, overall planning, direction, and technical and management oversight of NASA's science portfolio. This includes focusing on the scientific exploration of Earth, the Sun, the solar system, and the universe. I worked closely with the Mission Directorate Associate Administrator and often represented him in various capacities.

My management style was more about leadership than just managing. I met with Division Directors on a weekly basis and gave them the freedom and responsibility to run their Divisions within certain guidelines, consistent with the freedom to manage with accountability. I believe in treating all staff, including program managers, program executives, scientists, engineers, support functions, and administrative staff, with respect, even though their roles may seem more or less important at times.

I like to start my day early, giving myself an hour of uninterrupted time to review tasks, prepare for the day, and address any issues from the previous night. One morning, while reading my emails, my administrative assistant surprised me by coming to my office, closing the door, and, out of the blue, asking if I was a Christian. It caught me off guard, but it was definitely a memorable moment.

I asked her what had prompted her to ask that question, and she explained that she and the other administrative assistants were chatting. They all agreed that I must be a Christian because I treated them with respect, just like I do the scientists and engineers. They said they hadn't experienced that with other bosses. It was nice to hear that I was "walking the talk" of trying to treat everyone with respect. I didn't just settle for mediocrity, but I made sure to treat everyone fairly.

Accountability is crucial for the success of any mission. One challenging situation I faced was when my boss and I decided to remove someone from a key position within the Mission Directorate for various performance reasons. We followed all the necessary legal procedures but ended up being sued for the removal. I vividly remember the stress and tension while sitting in the courtroom, explaining our decision. After enduring four long hours on the stand,

I finally finished my part and headed straight to our cabin in West Virginia. I decided to build a lower deck to help release some of the stress I was feeling. Surprisingly, I may hold the record for the fastest deck construction ever. Who knows, but it definitely helped me unwind.

I was privileged to co-chair monthly program reviews between NASA and the National Oceanic and Atmospheric Administration (NOAA) for missions that were key to early weather predictions and monitoring.

Launching science satellites and rovers was a major task for NASA's Launch Service Provider (LSP). I collaborated with the LSP to ensure clear communication, particularly regarding any launch vehicle issues that might impact the science payloads. They ensured both the satellite and the launch vehicle were ready.

Sitting at the launch control monitoring stations, I couldn't help but think about all the things that could go wrong. I was familiar with every step of the launch process and held my breath as the launch vehicle progressed through each phase. Things like MaxQ, when the rocket hits its maximum stress point, always made me a little nervous. Even though launches may seem simple, they're actually pretty risky, and things could definitely go south.

The Washington Post captured our excitement as we learned that the JUNO spacecraft had successfully entered Jupiter's orbit. Entering the orbit at the incorrect angle or speed could have been disastrous, resulting in either skipping away or crashing into Jupiter. I turned to

Scott, the Principal Investigator of JUNO, and, somewhat speechless, only managed to say, "Wow, now for the real science."

Image credit: NASA/JPL

This was just one of the many thrilling experiences associated with NASA missions. Per the JPL press kit, *"The principal goal of NASA's Juno mission is to understand the origin and evolution of Jupiter. Underneath its dense cloud cover, Jupiter safeguards secrets about the fundamental processes and conditions that governed our solar system during its formation. As our primary example of a giant planet, Jupiter can also provide critical knowledge for understanding the planetary systems being discovered around other stars.*

With its suite of science instruments, Juno will investigate the existence of a possible solid planetary core, map Jupiter's intense magnetic field, measure the amount of water and ammonia in the deep atmosphere, and observe the planet's auroras.

Juno will let us take a giant step forward in our understanding of how giant planets form and the role these titans played in putting together the rest of the solar system."

Each mission launch carried its own set of excitement, expectations, thrills, challenges, and anxious moments. I had to prepare two speeches for each launch—one for a successful launch and one for if something went wrong. Luckily, throughout my SMD career, I only had to use the "We had a successful launch" speech. It was a relief every time I got to deliver that good news to the press.

A special memory was when Dr. Jim Green, the Planetary Systems Division Director, and I traveled to the European Space Agency (ESA) auditorium for a joint ESA/NASA celebration of the Rosetta probe landing on the Jupiter-family comet 67P/Churyumov-Gerasimenko. The mission design was to chase, orbit, and land on a comet. On 12 November 2014, Rosetta's lander Philae was deployed to the surface. Philae carried a suite of instruments for imaging and sampling the comet nucleus. The Rosetta mission excited us with its scientific aspects. On the auditorium stage at ESA, we discussed the mission's achievements and awaited the end of the lander portion of the mission.

We watched the clock, feeling a mix of bittersweet emotions as the final moments approached. Within an hour of the expected last communication from the lander, anticipation grew. At 30 minutes and then 15 minutes, we still received signals. Finally, a signal confirmed the lander had completed its mission, followed by silence. After 15 minutes without further signals, we declared the lander portion of the mission over. The Rosetta orbiter continued to track the comet through perihelion, examining its behavior before, during, and after the probe landing.

In a moment of stunned silence, I turned to Jim and compared the experience to being in the room when my mother passed away. Though not the same, the attachment to the mission was strong, and the ending brought unexpected emotions.

An Unexpected Interruption

Life-changing events can shift your perspective. In August 2015, I was at a two-day Agency workshop when I started feeling my leg swelling up and developed some severe back pain. I brushed it off, thinking it was just a pinched nerve.

The next morning, while getting ready for the workshop, I was struggling because my back pain was so intense that I could barely walk. My wife noticed and insisted I see a doctor. I tried my best to hide my pain, but this time, it was obvious beyond what I could hide. Since our family doctor was out of the country, we ended up at the 24-hour emergency care center. I told the doctor about my pinched nerve theory, but he chuckled and said we needed to do a thorough evaluation to pinpoint the cause of the pain. I think he already had a hunch.

After a CAT scan and X-ray, reality hit me hard. I was wheeled back to my room, hooked up to probes, and given a shot in the stomach. It wasn't until I saw my wife's face turning ghostly white that I started to worry. The doctor informed me of several clot issues and asked which hospital I preferred, saying the ambulance was ready. At that point, I had no comprehension of how serious the blood clots were. Instead, my experience working on the farm with machinery and my engineering background working with space flight hardware had me focusing on the mechanics of the ambulance rather than the health crisis during the bumpy ambulance ride to Inova in Alexandria, VA.

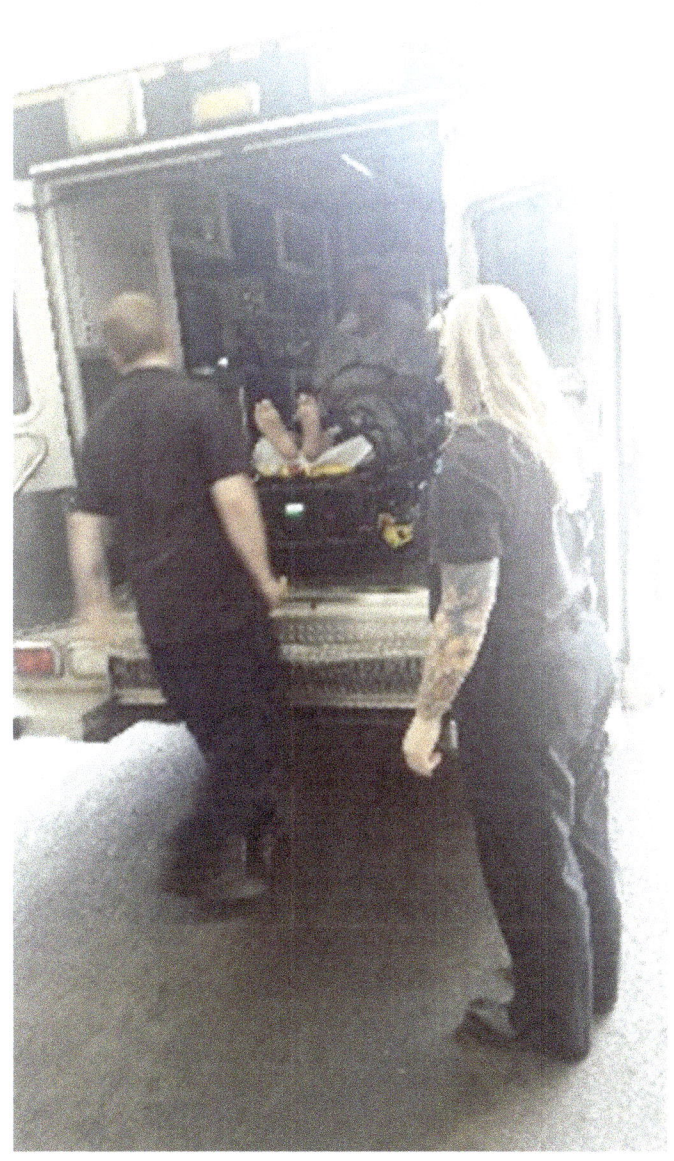

Me being transported to the hospital.

In the ICU, the doctors explained the plan to tackle the clots. It turns out that I had a solid clot from my hip to my ankle and massive clots in both lungs. It was hard for me to grasp the seriousness of it all fully.

My 5-year-old grandson came to visit me with a bear with a sweet message: "No day is so bad that a bear hug can't fix it."

My grandson surprised me with this teddy bear in the hospital.

Pressing the bear's belly button caused the bear to play the song "So You Had a Bad Day." He would adjust my bed to make me more comfortable or to ease his own nerves. Either way, it temporarily took my mind away from the blood clots. I still have that bear today, as it's a constant reminder of life's frailties. After three days and several procedures to deal with the clots in my leg, I was finally able to go home. After two weeks of rest, I was given the green light to return to work part-time until I fully recovered.

I was surprised that I didn't receive many phone calls from work while recovering at home. It turns out that my administrative assistant instructed my staff not to contact me and that she and Lauretta were in constant touch. They would let me know if any issues needed my attention. I had complete faith in my staff to keep things running smoothly in my absence. As I mentioned earlier, my goal was to be a leader, not a micromanager. The staff knew what to do and carried out their tasks perfectly.

Even though I didn't fully understand the seriousness of the blood clots, I could see the worry in others and started to think about retiring at the age of 56 rather than at 65, as is most common.

In hindsight, I realized there were early signs of blood clots that I had ignored. While laying pavers at our West Virginia cabin, I felt short of breath and attributed it to dust. The shortness of breath persisted until my breathing capacity dropped to about 30%. Alone and concerned, I feared something serious resting against the wall in the event I would pass away. I didn't want Lauretta to find me lying on the pavers. I experienced severe headaches, later identified as micro strokes, before my leg swelled and severe back pain set in. My sister, an RN, worried for my life, noting her friend's recent death from blood clot complications. The lesson: Don't ignore health symptoms.

Cabin in WV, where I installed pavers and rested against the wall.

Acting Mission Directorate Associate Administrator (MDAA)

Although I had already decided to retire from NASA when the SMDAA announced his retirement, I agreed to step in as the Acting Mission Directorate Associate Administrator until a permanent replacement was identified. Once NASA picked someone for the job, I stuck around for a few more months to help with the transition.

From the NASA website, "The Office of the Associate Administrator for NASA's Science Mission Directorate (SMD) is responsible for directing and overseeing the nation's space research program in Earth and space science. The Directorate engages the external and internal science community to define and prioritize science questions and seeks to expand the frontiers of five broad scientific pursuits: Earth Science, Planetary Science, Biological and Physical Sciences, Heliophysics, and Astrophysics.

Through a variety of robotic observatory and explorer craft, and through sponsored research, the Directorate provides virtual human access to the farthest reaches of space and time, as well as practical information about changes on our home planet."

As the Deputy MDAA, I was ready for the role. I was honored to receive the keys to the Science Mission Directorate from my boss, Dr. John Grunsfeld, an astronaut known for his five Hubble Space Telescope repair and upgrade missions. Working with John taught me the value of detail and calculated risks. When I asked about the risks he took, he explained that thorough training and understanding made the missions worthwhile despite the dangers.

Photo credit NASA: Figurative key to the MD handover ceremony.

The Science Fleet represents the fleet within the SMD Mission Directorate (MD). When I retired in November 2016, the MD was responsible for an annual budget of approximately $5.6 billion, encompassing 114 missions at various stages of development and operation.

The SMD MDAA was responsible for providing direction, oversight and chartering the path forward for new missions. The majority of the missions were centered around Astrophysics, Heliophysics, Earth Science, and Planetary Science, each within their respective Division. Each opportunity was unique, building upon scientific discoveries from previous missions and aligned with the National Academy of Science Decadal recommendations.

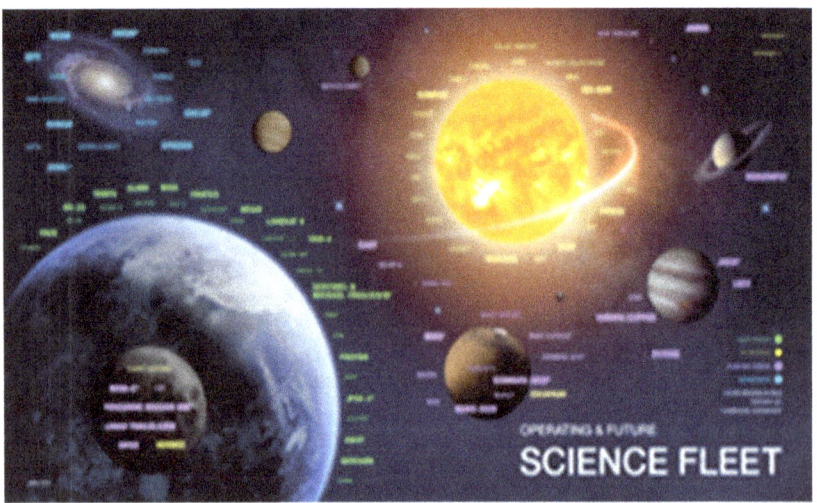

Image credit NASA SMD:
Science Fleet representing all the SMD Science Missions

One mission to Pluto, launched nine years earlier, required a critical decision about navigating through a band of cosmic particles. This posed a risk of damaging the instruments or potentially compromising the mission by maneuvering around the particles and missing significant scientific opportunities. As the Acting MDAA, I was aware that a wrong decision could lead to intense scrutiny by external stakeholders, including the White House and Congressional staff. The Principal Investigator convened multiple meetings with scientists and engineers to assess the risk based on new information from the approaching spacecraft. Ultimately, after analyzing the findings and recommendations, the PI and I decided to proceed through the dust, weighing the scientific benefits as greater than the risk of damage to the spacecraft. This decision-making process exemplified the collaborative working relationship between the MDAA and Principal Investigators on various missions.

The Mars Atmosphere and Volatile EvolutioN (MAVEN) mission faced significant challenges as we approached the launch window. This mission was designed to determine how much of the Martian atmosphere has been lost over time by measuring the current rate of escape into space and gathering sufficient information on relevant processes to extrapolate backward in time. Additionally, MAVEN was equipped with a backup communication capability for the Mars rovers.

Mars missions have specific launch windows, and for MAVEN, missing this launch window would mean waiting another 18 months for the next opportunity, resulting in additional costs. Initially, everything was on track for the first launch opportunity, but there was an issue: the government had entered a shutdown, MAVEN was deemed non-critical, and a stand-down order was issued.

This classification meant we would likely miss the launch window, incurring further expenses while waiting for the next opportunity eighteen months later. One member of my staff proposed a brilliant idea to justify reinstating the authorization to continue to work on the MAVEN mission. Existing communication relays for the Mars rovers were beyond their expected lifespans, and failure could result in a loss of communication with the rovers. Therefore, we justified the necessity of launching MAVEN by highlighting the need for its backup communication capability in case the aging relay satellites failed and we lost communications with the Rovers on the Martian surface, potentially ending the rover mission. We believed this would designate the MAVEN mission as a critical mission.

We had to coordinate this rationale with the implementing center, the NASA Legislative Affairs Office, the White House Office of Management and Budget (OMB), and the Office of Science and Technology Policy. Ultimately, MAVEN was successfully launched at the first available opportunity.

Photo Credit United Launch Alliance:
Standing in front of the MAVEN, prepared for launch

Another significant mission was the study of the Sun, known as Solar Probe Plus, later renamed Parker Solar Probe, which was developed at the Johns Hopkins Applied Physics Laboratory in Laurel, MD. During my tenure as MDAA, I had the privilege of observing its development, advocating for continued funding, and monitoring its progress through various reviews.

A particular area of interest was the heat shield, which required innovation to withstand the immense heat from the Sun and protect the spacecraft's critical systems. For context, the Sun is 93 million miles from Earth, and the Probe was designed to approach within 3.8 million miles of the Sun's surface in the corona region, enduring temperatures up to 2,500 degrees Fahrenheit.

I had many meetings at the Executive Office of the President (EOP) to update presidential advisors and others, as well as visits to Capitol Hill to update staffers and members of Congress.

A notable meeting took place with Representative John Culberson, during which we discussed the Mars rover and its accompanying helicopter. We reviewed the features of the helicopter and reached an agreement on a critical aspect of the mission. It was imperative for me to classify the helicopter as a technology demonstration rather than a mission-critical component, as was the designation of the rover itself. This classification ensured that the primary mission, centered on the rover itself, would not be considered a failure in the event of a premature failure of the technology demonstration helicopter. Representative Culberson concurred, and the directive to NASA reflected this distinction.

Another trip to Capitol Hill was to explain why we had to cancel a mission. My boss and I had to justify the decision. I remember sitting in the conference room with the staffers, discussing the challenges. Halfway through, I asked a simple question: Should we focus on keeping costs down like we've been told, or should we just let missions go over budget and move on? The conversation shifted to providing more information rather than questioning our choice.

I loved giving briefings to members of Congress and their staff on different topics. One meeting of interest was an educational briefing to Congressional members, staff, and other colleagues about key elements of Earth Science.

Briefing Congressional members, staffers,
and families about Earth Science

Final Launch Experience as a NASA Employee

My last science mission as Acting MDAA was a big one—we sent a satellite to collect samples from the Bennu asteroid and bring them back to Earth as part of the OSIRIS-REx mission. It was an honor to have family and friends with me for my final launch, but the feeling was surreal.

Photo Credit: NASA. O-REX mission as my last launch as a NASA official

"This mission exemplifies our nation's quest to boldly go and study our solar system and beyond to better understand the universe and our place in it," said Geoff Yoder, Acting Associate Administrator for the agency's Science Mission Directorate in Washington. "NASA science is the greatest engine of scientific discovery on the planet and OSIRIS-REx embodies our directorate's goal to innovate, explore, discover, and inspire."

As I gave my final overview talk to guests at the OSPII building at the Kennedy Space Center, I felt a mixture of emotions. The excitement and pride of sharing this moment with loved ones were tempered by the bittersweet realization that this would be my last launch. After the talk, I rushed to the launch control site for the last time, recalling the many trips from OSBII to the control center several miles away.

Memories of similar experiences during launches from the Vandenberg Air Force Base flooded my mind. The heavy fog in the mornings would often obscure the rocket shortly after launch, adding a layer of mystery and anticipation. Each launch was a unique experience, and this final one was no different.

About 30 seconds prior to liftoff, a group of us went to the roof of the building to watch and experience the launch. As the countdown reached zero, the flames at ignition lit up the sky, followed by the powerful rumble of the engines. The sheer force of the rocket's ascent was awe-inspiring, and shortly thereafter, we felt the soundwaves reverberate against our bodies. It was a sensation that never got old, a reminder of the immense power and precision required for space exploration.

Watching the rocket soar into the sky, I couldn't help but reflect on the countless hours of preparation, teamwork, and dedication that brought us to this moment. Each launch represented a culmination of effort and innovation, a testament to what we can achieve when we work together toward a common goal.

As the rocket disappeared into the distance, I felt a profound sense of accomplishment and gratitude. This final launch was not just an end but a celebration of a journey filled with challenges, triumphs, and unforgettable memories. The support of my family, friends, and colleagues made it all the more meaningful.

In the end, every launch is a new experience, a new chapter in the story of space exploration. This final chapter of my career with NASA was a fitting tribute to the adventure and wonder of spaceflight, a journey that I will carry with me for the rest of my life.

Retirement from NASA

Upon my retirement from NASA, former NASA Administrator General Charles F Bolden Jr. commented, "With more than 16 years in the industry and 16 years at NASA, Geoff's story is not only of individual success and hard work, but also of NASA's transition to a new era of space exploration, in which he played many key leadership roles. He has accomplished what most of us come here hoping to do—move our mission—and America's space program—forward."

My retirement party was truly unforgettable, with my wife and son by my side. Throughout my career, I have always tried to uphold high ethical standards and treat everyone with respect, regardless of their position.

As the speakers took to the podium to share some words and raise a toast, the focus was not on the missions I had been a part of but rather on how I maintained high standards and showed care and respect for my team. It was the perfect send-off.

I was touched as I received various tokens of appreciation for my time at NASA. Congressman Chris Van Hollen even went above and beyond by flying a United States flag over the U.S. Capitol in my honor, recognizing my service to NASA and the country.

Certificate stating the United States Flag was flown over the US Capital in my
honor. Presented by the Honorable Chris Van Hollen, Member of Congress

Not everything was focused on integrity or mission work. Jim
Norman, who was in charge of launch services, added humor by
sharing his fourteen favorite Country songs.

#14: Let's Do Something Cheap and Superficial!
From the Smokey and the Bandit 2 soundtrack, performed by Burt
Reynolds and written by Richard Levinson (BMI)
#13: Her Only Bad Habit was Me! From the BMI database
#12: You Ain't Much Fun Since I Quit Drinkin' By Toby Keith (BMI)
#11: How Can You Believe Me When I Say I Love You, When You
Know I've Been a Liar All My Life? By Burton and & Alan Jay Lerner
9ASCAP) for the film Royan Wedding
#10: She Got the Ring, I Got the Finger By Chuck Mead & His
Grassy Knoll Boys

#9: I'm Still Missin' You Baby, But My Aim is Getting' Better! By Greg Perkins/Monty Holmes

#8: If She Hadn't Been So Good Lookin' I Might Have Seen the Train By Anonymous

#7: I would Kiss You Through the Screen door, But I Would Strain Our Love Unattributed

#6: Meet me in the Gravel Pit Honey Cauz I'm a Little Boulder There! Unclaimed

#5: Mommy, Can I Still Call Him Daddy? By Dottie & Bill West

#4: My Wife Ran Off With My Best Friend, and I Sure Do Miss Him By Phil Earhart (BMI)

#3: If I'd Shot You When I wanted to I'd be out of Jail by now! By Ruben Darnell

#2: Get Off the Table, Mable, The Two Dollars is for the Beer! By Bull Moose Jackson

#1: I'm so Miserable Without You it's like You're Still Here By Mike Haduk

I received a picture of the weather map from the day I was born in Meyersdale, PA, from colleagues at NOAA, along with a written document detailing the weather on that day. This reminded me of the numerous joint meetings that I co-chaired with my counterparts at NOAA. These reviews were an important aspect of improving the nation's weather forecasting capabilities.

I was honored to receive a Mission Portfolio book featuring various missions that I was involved with throughout my career, with comments from scientists, engineers, administrative staff, and others. This book is something I will always cherish.

Astronaut and Kennedy Space Center Director Robert Cabana gave me an autographed plaque with a flag that flew on Space Shuttle

STS-135. I worked with Mr. Cabana on various occasions, and it was an honor to receive this plaque remembering the times and interactions with the team at the Kennedy Space Center.

I received other plaques from the Johnson Space Center and NASA Centers, but two in particular are extremely special to me. One is from NASA Administrator General Charles Bolden Jr. and represents all the mission pins associated with the Space Shuttle from its inception to its retirement. The other plaque represents the missions to the International Space Station (ISS) from its inception until the day I retired from NASA.

Adding humor to my retirement, the Astrophysics Division created a bobblehead doll as a parting gift. Sometimes, it's the little things that make a difference.

CHAPTER 9:
A NEW JOURNEY

Retirement isn't all about sitting around and being bored. A few months after leaving NASA in November 2016, I started working at the John Hopkins Applied Physics Lab as a part-time staff member. My job involves supporting independent investigations, reviewing proposals, and offering general advice.

When you hold a high-level government position, I think it's important to give back to the community. I love Science, Technology, Engineering, and Math plus Arts and Design (STEM+AD) education, so I've been talking to middle and high school students in classrooms and assemblies. I remember speaking to a group of 5th to 7th graders as a special guest for their afterschool program. I shared my high school experiences and how I made it to NASA, and most of the students were inspired and thought they could achieve the same type of success, even if it were in a different field. But one young girl raised her hand and said she didn't think she could succeed. It broke my heart to see her doubt herself, reminding me of my own struggles in high school.

During the COVID pandemic, I instructed a semester of Aerospace Engineering to junior and senior high school students. Most classes were conducted online, with key topics discussed in person. I invited guest speakers, including a Project Manager and Program Scientist who shared their essential skills and real-time examples relevant to their tasks, an astronaut who recounted his experiences in space, a Nobel Laureate who discussed his initial failures and eventual success through tenacity and perseverance, and a cyber security expert who spoke about her educational challenges and the significance of cyber security. Determined to provide a memorable experience despite the isolation of online learning, I was gratified by the heartwarming feedback received from the students at the end of the semester.

I assigned a paper exercise design project to build a Lunar hopper, focusing on objectives and design requirements, including environmental constraints. I left room for creativity but included one unattainable requirement that over-constrained the mass limits without the students knowing the implications of the added requirement. The students only discovered this challenge during our final in-person lesson. When they realized their design couldn't meet the requirements, we discussed real-life issues of unmet performance expectations and the need to reevaluate constraints. This "fail without failing" exercise was inspired by a discussion with astronaut Kathy Sullivan, who emphasized the importance of persistence despite setbacks. At the end of the semester, students said this was the most memorable lesson.

This experience has deepened my respect for educators. I witnessed firsthand that teaching demands significantly more time than is typically allotted for a single class.

As this initiative gained recognition, I was invited to speak at career day events for middle and high school students at various institutions. Additionally, I became a member of the local school district's Citizen Financial Advisory Group, where I contribute my expertise in financial management.

Being invited to speak at these career events has been particularly rewarding. It has allowed me to give back to the community and inspire the next generation.

Even after retiring, I found myself with some free time, so at the urging of residents, I decided to become a Town Commissioner. It was a great way to use my knowledge to benefit the town.

In 2019, NASA asked me to lead the independent review team for their VIPER rover mission to the Moon. The rover was an incredible opportunity to provide valuable scientific input to the community and the world. About the size of a golf cart, NASA's Volatiles Investigating Polar Exploration Rover, or VIPER, is ideal for traversing the Moon's South Pole. As NASA's first mobile robotic mission to the Moon, VIPER would directly analyze ice on the surface and subsurface of the Moon at varying depths and temperature conditions within four main soil environments. The data VIPER transmits back to Earth would be used to create resource maps, helping scientists determine the location and concentration of ice on the Moon and the forms it's in, such as ice crystals or molecules chemically bound to other materials. As it prospects for ice from the surface, VIPER's findings will inform future landing sites under Artemis by helping to determine locations where water and other resources can be harvested to support a long-term presence on the Moon. Bringing everything we need from Earth for long-term exploration in space would be very costly, so using resources found on the Moon, like water, could be a game-changer for human space exploration to the Moon, Mars, and beyond.

To access the most promising ice reserves, VIPER must maneuver a highly cratered surface with soils of varying levels of compaction. The rover's components are rugged enough to withstand the encroachment of lunar dust, blasts of cosmic rays, and dramatic swings in lighting and temperature. With its onboard suite of instruments to find out what lurks below the lunar surface, VIPER will help inform our understanding of how to live and achieve a long-term presence on the lunar surface.

Despite the promising scientific benefits to humanity, NASA canceled the VIPER project in 2024 due to overall funding issues, including that of the lander. The review team was disbanded in 2024 after the rover successfully completed critical testing.

At the same time, I was also co-leading an independent review of NASA's Earth Observatory System plans. This review included assessing how proposed earth science satellites work together in unison to perform a transparent observatory to end users.

It may seem like I never really retired! Some of these activities started during the COVID-19 pandemic, so most of the work could be done remotely.

One of my priorities after retiring from NASA was to spend more time with my family. Lauretta and I even went to Las Vegas to see the Blue Man Group. It was such an exciting adventure! A friend in Vegas arranged the tickets for us, but there was a special surprise in store.

Blue Man Group presented us with autographed (blue lip print) drum sticks

We were told to arrive early for the show and were seated halfway back in the auditorium with an aisleway directly behind us. We were confused, wondering why our friend's connections didn't get us better seats. But then, just before the show started, an usher came to us and asked us to follow him. As we walked across the aisle, a spotlight shone on us, and the music played, "You're late, you're late, you're holding up the show, you're late." It was embarrassing, to say the least! We were then seated in the center, eight rows from the front. Later, we found out that those closer than eight rows would get splattered with water during the show.

After the show, an usher kindly gifted us a pair of drumsticks that were used during the performance. One of the members of the Blue Man Group even autographed the sticks for us, leaving his signature blue lip prints on them.

Lauretta and I have been on quite the adventure since then, exploring different venues and taking trips to places I never had time for before. We visited Paris, cruised the Alaskan inner passage with family to celebrate my retirement, and later went on the same cruise with my son, daughter-in-law, and grandson.

The cruise bug bit us, and we ended up cruising the Panama Canal, exploring the Hawaiian islands, and going on various Caribbean cruises.

Our grandson, Bryton, is an avid hockey enthusiast and an active player in various leagues, achieving notable successes throughout his young career. As proud grandparents, we make it a priority to attend all of his games. We aspire to see him play for a semi-professional team or even in the NHL in the future.

Our adventures are far from over. I've decided to dedicate more of my time to leisure activities and minimize accepting requests for studies or other non-educational activities.

My involvement with NASA and my experiences will always be significant as I continue to advocate for the benefits NASA provides to society. I had the opportunity to host Delaware Congressman Brian Pettyjohn at the Wallops Flight Facility (WFF) for an Antares launch event, which aimed to send the Cygnus spacecraft to the International Space Station (ISS). Though the launch was delayed due to a software issue, Brian attended the successful launch the following night. Additionally, I hosted approximately 20 senior citizens at a previous Antares launch at WFF. These events were noteworthy, especially since the WFF Center Director previously worked under my leadership in the Science Mission Directorate before accepting the Center Director position and working with Congressman Pettyjohn when I was a Town Commissioner.

CHAPTER 10:
REFLECTIONS ON MY CAREER

Throughout my career, I've seen how the influences we experience in our early years can really stick with us, impacting our decisions and choices, whether they're positive or negative. I relate to the many young people I have met along the way in that I, too, felt as if I was not good enough to achieve anything and felt very alone. Yet, overcoming what others had said about me when I was young, I took from the many adventures I have shared and am now determined to pass this knowledge on to students to encourage them to pursue their dreams and not be held back by others' opinions of them.

Growing up on a farm taught me some important life lessons about honesty, integrity, and responsibility. Those early lessons have shaped my career and still impact the decisions I make today. Even building model rockets as a kid taught me about following rules and understanding the consequences of being careless.

I've always believed in finding the good in people, giving encouragement when needed, and really thinking things through before making a move. Patience, faith, and leadership have always been my guiding principles. Treating others with respect is crucial, but sometimes you have to face challenges head-on to grow.

Seeing the people I've worked with or mentored succeed has been so fulfilling. Keeping my integrity intact and dealing with problems directly have always been my top priorities in any job I've had. Consistency in actions is key, especially since others are always watching.

My many jobs at NASA may appear like I couldn't keep a job, but the directed promotions benefited the Agency. Each job led to more visibility and critical roles within the Agency.

I'm thankful for the coworker who pushed me to go to community college, which was the start toward my NASA career, and I will forever be grateful for him taking the time to encourage me in ways that altered the journey. I am sure Lloyd is not aware of the impact he made on me. We all touch someone's life without knowing the impact we have on them.

I wish there had been more support for me during my high school years. That's why I'm committed to mentoring and supporting students now, to help them avoid some of the same struggles I experienced.

Who would have thought that a farm kid from a small town in Pennsylvania could rise to lead the largest civilian space science organization in the world as a Senior Executive Service member of the United States of America? One way to summarize my lessons learned experience is to follow the Golden Rule: Do Unto Others As You Would Have Them Do to You.

A final consideration is knowing when to step down. My blood clot health scare prompted me to reflect on the future. As I considered the events over the years, I realized that if I had succumbed to the blood clots, NASA would likely have paid tribute to my contributions and promptly initiated a replacement process. That's how it should be. However, I wanted to share my experiences in a way that would encourage those who follow not to give up. It became evident that it was necessary to allocate more time to my family and to provide mentorship to students.

The Hubble images, which employ both optical and infrared technology to capture the Carina Nebula, underscore the importance of these observations. These images suggest that our initial perceptions may not encompass the entirety of what exists. As illustrated by the Hubble photographs, there is frequently more information than initially apparent; therefore, it is advisable to refrain from drawing premature conclusions. There is often additional information that extends beyond surface observations.

Image Credit: NASA, ESA, and M. Livio and the
Hubble 20th Anniversary Team (STScI)

None of my successes would have been possible without the unwavering support of my wife, Lauretta, for whom I am deeply grateful for her encouragement and companionship.

Principles to Live By:

Before I joined NASA, I had already learned some important life lessons—but in the fast-paced world of space exploration, it was easy to lose sight of those principles. To keep myself grounded, I created a one-page list of sayings and placed it at the base of my computer monitor. I'm not sure who originally came up with many of them, but over time, they became personal mantras—reminders to stay present, be grateful, and not sweat the small stuff.

Key Leadership Attributes to Follow (*the list I had at my computer monitor*)

Need new goals for each new day
- Sense of purpose
- Spirit of adventure
- Capacity for growth

Develop characteristics
- Uncompromising integrity
- Working priorities (not only developing them)
  - *Blinding flash of the obvious*
    - Start with number one on the priority list and continue doing priorities
    - Expression: if you have a frog to swallow, don't look at it too long, if you have two, swallow the biggest one first
- Courageous – either you or your fear takes control
  - Expression: Don't wait until all the stop lights are green before going across the city
- Start gaining altitude before making a turn
- A crisis must never be experienced a second time
- Goal orientation: ability to make the tough decision
- Don't get frustrated: sometimes change is needed
  - Expression: can't teach a pig to sing, wastes your time and it makes the pig mad trying
- Inspired enthusiasm
- Level headed – good speed in grasping the facts
- Ability to select and develop good people who often better than you
- There is no situation so bad that a single action can not make it worse

4 key works
- Dream
- Study everything you get your hands on
- Plan you time, time your plan (its ok to have a type by the tail if you know what to do next)
- May you never be less than your dreams

5 to 1 rule
- provide a minimum of 5 compliments to 1 criticism

3 out of 5 rule
- On the average: 3 out of 5 days a week, feel good about what was accomplished, believe you contributed to the mission, and will return the next week. Anything short of "3 out of 5" indicates a wrong job assignment or in the wrong position

Keep your head above the clouds and your feet firmly planted on the ground – then move step by step.

One-page list of sayings that I kept by my computer

When I first started at JSC, I also began jotting down key events and meaningful phrases that had shaped my journey—lessons learned, moments of clarity, and advice from mentors. It was like taking a trip down memory lane, and each entry brought with it a renewed sense of direction and calm. Whether it was "sometimes change is needed" or "don't wait until all the stop lights are green before crossing the city," the sayings reminded me that life has its ups and downs, but everything ultimately works out.

Most importantly, I always came back to one core value: "Maintain uncompromising integrity."

Key Leadership Attributes to Follow (the list I had on my computer monitor)
- **Need new goals for each new day**
  - Sense of purpose
  - Spirit of adventure
  - Capacity for growth

- **Develop characteristics**
  - Uncompromising integrity
  - Working priorities (not only developing them)
  - Blinding flash of the obvious
  - Start with number one on the priority list and continue on priorities
  - Expression: If you have a frog to swallow, don't look at it too long. If you have two, swallow the biggest one first.
  - Courageous—either you or your fear takes control.
  - Expression: Don't wait until all the stoplights are green before going across the city.
  - Start gaining altitude before making a turn
  - A crisis must never be experienced a second time
  - Goal orientation: ability to make the tough decision
  - Don't get frustrated—sometimes change is needed
  - Expression: You can't teach a pig to sing; it wastes your time, and it makes the pig mad trying
  - Inspired enthusiasm
  - Levelheaded—great speed in grasping the facts
  - Ability to select and develop good people—hire ones better than you
  - There is no situation so bad that a single action can not make it worse.

- **4 Key Themes**
  - Dream
  - Study everything you get your hands on
  - Plan your time, time your plan (it's ok to have a tiger by the tail if you know what to do next)
  - May you never be less than your dreams

- **5 to 1 Rule**
  - Provide a minimum of 5 compliments to 1 criticism.

- **3 Out of 5 Rule**
  - On average, 3 out of 5 days a week, feel good about what was accomplished, believe you contributed to the mission, and want to return the next week. Anything short of "3 out of 5" indicates a wrong job assignment or being in the wrong position.

- **Keep your head above the clouds and your feet firmly planted on the ground—then move step by step.**

Key Life Lessons:

My life experiences are deeply meaningful to me—each moment, each person, and each lesson has left an unforgettable mark. I carry them close to my heart and will always cherish the memories that have shaped who I am. Reflecting on those moments now, I realize how important it is for educators to recognize and nurture potential leaders early in a child's life, just as my 5th-grade teacher did for me. That kind of belief and support can change the course of a life, and I'm forever grateful for the people who saw something in me before I saw it in myself.

And now, I leave you with the life lessons that these experiences have taught me, hoping that they inspire you to keep reaching for the stars or to encourage a young person in your life to do so.

Lesson #1: Focus on the good in people

One of the most important lessons life has taught me is to always look for the good in people, even when it seems hard to find. It's so easy to get caught up in pointing out someone's flaws or mistakes, but focusing on their positive qualities can really change your perspective. When you start looking for the good in others, you'll be amazed at how much more positive and uplifting your interactions become. It's about shifting your mindset from judgmental to empathetic and from pessimistic to optimistic. Remembering that everyone has their own struggles and strengths can help foster a sense of understanding and compassion toward others. We all make mistakes—focusing on the positive aspects of people rather than the worst is just one way we can create a more harmonious and supportive world for ourselves and those around us.

Lesson #2: Offer support and mentorship to those who need an extra boost.

Imagine you have a friend or colleague who has all this amazing potential just waiting to be unleashed but, for whatever reason, is not advancing. By offering your support, encouragement, and helpful advice, you can help them navigate the challenges and obstacles they may face along the way. So don't hesitate to lend a hand—together, you can help your friend achieve great things!

Lesson #3: Actions have consequences

Actions have consequences—good or bad so think about the consequences before making major decisions.

Lesson #4: Experiences shape the attitude

Life has a way of throwing us into moments that completely shift our perspective, often through profound pain and unexpected loss. My experiences taught me how to put my own grief aside to support others, ultimately shaping my understanding of resilience, the quiet strength found in community, and the enduring human capacity to carry on, even when the weight feels unbearable.

Sometimes, these tough times end up being a blessing in disguise. They push us to grow and evolve in ways we never imagined possible, but it's all part of the journey. It might just lead you to a better perspective on life in the end.

Lesson #5: Patience is a virtue

Being patient is key, especially when it comes to making major decisions. It can be tempting to act on emotions and make impulsive choices, but it's important to take a step back and consider all the facts before making a decision. Rushing into something without fully understanding the situation could have negative consequences. Sometimes, waiting a little longer for more information or seeking advice from others can lead to better outcomes. Your future self will thank you for taking the time to make an informed decision.

Lesson #6: Walk through open doors

Sometimes, in life, you just have to take a leap of faith. Sure, it can be scary not having all the details lined up perfectly before making a decision. Still, sometimes, that sense of uncertainty is what actually leads us to some amazing opportunities we would never have discovered otherwise. By being open to new experiences and willing to take a chance on something, even if everything isn't crystal clear, we allow ourselves room for growth and unexpected blessings. Who knows what incredible adventures could be waiting on the other side!

Lesson #7: Maintain a standard of high integrity

But the most important lesson is what I learned growing up on the farm. Take responsibility for your actions, honor your commitments, and conduct yourself with high integrity. It's all about holding ourselves accountable and taking responsibility for our mistakes. It means standing by our word and following through on our promises, regardless of the circumstances. Being a person of integrity means doing the right thing even when no one is watching and staying true to our values and beliefs. It's about being honest, ethical, and trustworthy in everything we do. So, whether it's showing up on time for a meeting or admitting when we've made a mistake, taking responsibility, honoring commitments, and maintaining high integrity are key ingredients to building strong relationships and earning the trust of others.

ACKNOWLEDGMENTS

I currently serve as a Senior Advisor to the Charles F. Bolden Group, which I believe represents the same goals that I uphold.

I have a deep appreciation for the leadership and impact of Inter Astra™, led by the Charles F. Bolden Group. Their mission—to expand equitable access, cultivate transformative leadership, and harness the power of space-related tools to improve life for humankind—reflects the values that have guided my own journey. Through their work, they are helping to shape a more just, inspired, and interconnected future, grounded in purpose and possibility.

REFERENCES

James Webb Space Telescope

The James Webb Space Telescope (JWST), launched on December 25, 2021, is the scientific successor to the Hubble Space Telescope and complements and extends the discoveries of the Hubble Space Telescope.

The JWST follows the Earth around the Sun at a location called the Earth-Sun L2 point, located 1 million miles from the Earth.

More information related to the JWST can be found at https://science.nasa.gov/mission/webb/

Juno

Juno was launched on August 5, 2011, traveling nearly five years and 1.7 billion miles, evading showers of the most punishing radiation outside the Sun before reaching Jupiter and entering Jupiter's orbit on July 5, 2016.

More information related to the Juno mission can be found at https://science.nasa.gov/mission/juno/#more-about-the-mission

Hubble Space Telescope

Hubble Space Telescope launch in 1990 has changed the fundamental understanding of the universe. The telescope experienced an anomaly where the imagery focus was not optimum, requiring correction. NASA conducted a special mission to the HST to correct the focus issue. A total of 5 repair and upgrade missions were conducted on the HST, resulting in

Imaginary from the HST that is nothing less than breathtaking, enabling scientific discoveries that could previously only be theorized.

More information related to the Hubble Space Telescope can be found at https://science.nasa.gov/mission/hubble/

Volatiles Investigating Polar Exploration Rover (VIPER)

The VIPER is designed to explore the relatively nearby but extreme environment of the Moon in search of ice and other potential resources. This mobile robot was slated to land at the South Pole of the Moon on a 100-day mission. The critical information it provides will teach us about the origin and distribution of water on the Moon and help determine how we can harvest the Moon's resources for future human space exploration. As of this writing, the VIPER mission was canceled.

More information related to the VIPER mission can be found at https://science.nasa.gov/mission/viper/

OSIRIS-REx

The mission was launched on September 8, 2016, traveling to a near-Earth asteroid named Bennu. One of the mission objectives was to collect a sample of rocks and dust from Bennu's surface and return the samples to Earth. The samples were collected on October 20, 2020, and delivered to Earth on September 24, 2023.

More information related to the OSIRIS-REx mission can be found at https://science.nasa.gov/mission/osiris-rex/

NASA SMD Science Fleet

More information related to the SMD fleet of mission can be found at

https://science.nasa.gov/science-missions/

Pluto New Horizons

New Horizons launched January 19, 2006, was the first spacecraft to explore Pluto up close, flying by the dwarf planet and its moons in 2015.

https://science.nasa.gov/mission/new-horizons/

ESA mission

More information related to the European Space Agency Rosetta (rendezvous with a comet) mission can be found at https://rosetta.esa.int

Senior Executive Service (SES)

Members of the SES serve in the key positions just below the top Presidential appointees. SES members are the major link between these appointees and the rest of the Federal workforce. They operate and oversee nearly every government activity in approximately 75 Federal agencies

More information related to SES can be found at
https://www.opm.gov/policy-data-oversight/senior-executive-service/

NASA Educational services

https://www.nasa.gov/learning-resources/

FOND MEMORIES

An honor to have my sister and her husband join Lauretta and me for my last launch as a NASA official – The OSIRIS-REx Mission

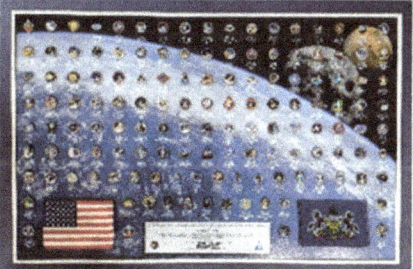

Mission Pins of all the Shuttle flights. The US and PA flag was flown on Orion Exploration Flight Test-1. The plaque is signed by NASA Administrator Charles F Bolden Jr.

Mission Pins for all flights to the International Space Station as of my retirement date. The NASA coin was Orion Exploration Flight Test-1. The plaque is signed by NASA Administrator Charles F Bolden Jr.

Mission Portfolio book presented to me by SMD,
representing missions with my involvement.

The inside cover of the Mission Portfolio book with well-wisher comments throughout the book.

The plaque presented to me was signed by the JSC Center Director, Astronaut Ellen Ochoa. The PA flag was flown on Space Shuttle Atlantis STS-135, traveling more than 5.2 million miles.

Credits: NASA, ESA, CSA, and STScI

One of the first JWST images is the Carina Nebula in all its glory, revealing the curtain of dust and gas and previously hidden baby stars. I was presented with this plaque of appreciation for my work on the JWST.

Plaque presented to me, signed by the KSC Director Astronaut Robert Cabana. The flag was flown on Space Shuttle Atlantis STS-135, traveling more than 5.2 million miles and completing 200 orbits of the Earth.

www.ingramcontent.com/pod-product-compliance
Lightning Source LLC
Chambersburg PA
CBHW070927130626
46555CB00001B/316